DATE DUE

APR 2 2 2012	
MAY 2 2 2012	

The Kuyper Lecture Series

The annual Kuyper Lecture is presented by the Center for Public Justice in cooperation with leading institutions throughout the country. The lecture seeks to enlarge public understanding of three dynamics at work in the world today: the driving influence of competing religions in public life, the comprehensive claims of Jesus Christ on the world, and the strength of the Christian community's international bonds. The lecture is named in honor of Abraham Kuyper (1837–1920), a leading Dutch Christian statesman, theologian, educator, and journalist.

The Center for Public Justice is an independent, non-profit organization that conducts public policy research and pursues civic education programs, such as the Kuyper Lecture, from the standpoint of a comprehensive Christian worldview. The Center advocates equal public treatment of all faiths and seeks political reforms to strengthen the diverse institutions of civil society.

Each book in the Kuyper Lecture series presents an annual Kuyper Lecture together with the responses given to it.

Books in the Kuyper Lecture Series

Mark A. Noll, *Adding Cross to Crown: The Political Significance of Christ's Passion,* with responses by James D. Bratt, Max L. Stackhouse, and James W. Skillen (the 1995 lecture).

Calvin B. DeWitt, *Caring for Creation: Responsible Stewardship of God's Handiwork,* with responses by Richard A. Baer Jr., Thomas Sieger Derr, and Vernon J. Ehlers (the 1996 lecture).

Caring for Creation

Responsible Stewardship of God's Handiwork

CALVIN B. DEWITT

with responses by

Richard A. Baer Jr.

Thomas Sieger Derr

Vernon J. Ehlers

edited by James W. Skillen and Luis E. Lugo

The
Center
for Public
Justice

P.O. Box 48368
Washington, D.C. 20002

 BakerBooks

A Division of Baker Book House Co
Grand Rapids, Michigan 49516

© 1998 by The Center for Public Justice

Published by Baker Books
a division of Baker Book House Company
P.O. Box 6287, Grand Rapids, MI 49516-6287

Printed in the United States of America

Library of Congress Cataloging-in-Publication Data

DeWitt, Calvin B.
    Caring for creation : responsible stewardship of God's handiwork / Calvin B. DeWitt ; with responses by Richard A. Baer Jr., Thomas Sieger Derr, Vernon J. Ehlers ; edited by Luis E. Lugo, James W. Skillen.
      p.   cm.
    Includes bibliographical references.
    ISBN 0-8010-5802-3 (pbk.)
    1. Creation. 2. Human ecology—religious aspects—Christianity. 3. Stewardship, Christian. I. Baer, Richard A. II. Derr, Thomas, Sieger, 1931– . III. Ehlers, Vernon J. IV. Lugo, Luis E., 1951– . V. Skillen, James W. VI. Title.
    BT69.D47 1998
    261.8'362—dc21   97-39356

For current information about all releases from Baker Book House, visit our web site:
                    http://www.bakerbooks.com

Contents

Foreword

James W. Skillen

The question of environmental stewardship becomes increasingly important with each passing year. No topic could be more appropriate for the annual Kuyper Lecture, by means of which the Center for Public Justice seeks to stimulate creative Christian thinking about the most important issues of our day. In the case of environmental stewardship, we face a truly global concern, one that should draw Christians together throughout the world to reflect, to pray, and to work together to do greater justice to God's creation.

We are pleased to present Calvin DeWitt's paper "Praising Rembrandt but Despising His Paintings"—which he presented at Fuller Theological Seminary on October 30, 1996—together with three responses. DeWitt is an evangel-

James W. Skillen (Ph.D., Duke University) is executive director of the Center for Public Justice. Previously he taught politics at Messiah, Gordon, and Dordt colleges. He is the author of several works, including *The Scattered Voice: Christians at Odds in the Public Square* (Grand Rapids: Zondervan, 1990) and *Recharging the American Experiment: Principled Pluralism for Genuine Civic Community* (Grand Rapids: Baker, 1994).

ical leader in the field of environmental studies, a professor at the University of Wisconsin at Madison, and director of the Au Sable Institute in Michigan, where people come from around the country and all over the world to study the environment. DeWitt paints a colorful picture in the pages that follow, urging us to take much more seriously the meaning of God's love for this world. He presents important religious principles that undergird responsible environmental stewardship.

DeWitt begins with what he calls a perplexing puzzle: All too often Christians who claim to love God show little interest in caring for the Creator's handiwork. In trying to understand that puzzle, DeWitt investigates seven degradations of creation for which humans are responsible, including soil erosion, deforestation, global toxification, and species extinction. Those who love the Creator ought to be resisting such degradation, and that is why DeWitt and others in the Evangelical Environmental Network took action in support of the Endangered Species Act (ESA) in 1995 and 1996.

The fact that not many Christians rallied behind the ESA is due to the power of another religion that dominates life in America, according to DeWitt. That is the religion of looking out for number one, the religion of self-centered consumerism. Christian faith stands directly opposed to this false religion because it calls us to seek first God's kingdom, not our own interests. God designed this world to work only if humans serve as good stewards of one another and of all other creatures, seeking God's will above all else. Human dominion over the earth must be a service-oriented, caretaking stewardship.

Within this framework, says DeWitt, we can answer three all-important questions of religious faith: Is Jesus Christ lord of creation? Is creation a lost cause? And, Whom are we following when we follow Jesus Christ? The answer to the first question is that Jesus Christ certainly is lord of creation, because according to the Bible, all things were created in him and through him and for him in the first place. Creation, then, is not a lost cause. To the contrary, God so loved the world that Christ came to redeem it, to reconcile it, to restore it. Those who follow Jesus, therefore, should be like Jesus. Jesus did not despise or disregard creation but rather came as a servant to heal and redeem it. Human dominion over the earth should be like Christ's dominion: a servant-like stewardship. When we follow Jesus, we follow a caring servant.

Looking at the world from this point of view, says DeWitt, we will be able to understand God's message to Job about Behemoth—the huge, unsightly hippopotamus. God loves this creature, and even though humans might not find it worthy of preservation, God says, "Keep your hands off my hippo!" If that is true of the hippopotamus, it is also true of God's creation generally. Humans may not destroy or degrade the creation at will but should act as caretakers and reconcilers. Christians really do have good news for creation, the good news that Jesus Christ is Creator, Author, Integrator, and Harmonizer of all things.

Richard A. Baer Jr., Thomas Sieger Derr, and Vernon J. Ehlers all express appreciation for DeWitt's work, yet each in his own way asks some of the critical questions that Christians must ask and try to answer if they are to be faithful stewards of the earth. Baer is professor of environmental ethics at Cornell University; Derr is professor of religion at Smith College; Ehlers—a physicist who taught at Calvin College for many years—now serves in Congress as the representative from Michigan's third district.

Professor Baer is delighted that DeWitt thinks so much about God, but he fears that DeWitt turns too quickly from God to his convictions about species-preservation policy. Evidence suggests that 99 percent of all extinct species disappeared before humans inhabited the earth, so what does that say about God's concern for species preservation? Baer agrees that modern secular ideologies are dangerous to Christian faith, but he believes that the challenge we face is to decide how governments, markets, schools, and industries each need to accept appropriate responsibilities to care for creation, particularly for human beings. "My comments," says Baer, "should be understood more as a word of caution than as a deep disagreement with him." But we do need to be "wary of those forms of environmentalism that treat nature as ultimate reality, and human beings as a blight on this fragile planet."

Dr. Derr expresses some of the same cautions about DeWitt's presentation and urges us to pay close attention to the complexity of the judgments that must be made about environmental degradation. Derr remembers, for example, when the great environmental concern was about global cooling. Now it is about global warming. Have we listened carefully enough to all the evidence? Furthermore, we need to be careful not to let our awareness of environmental problems lead us automatically to one political approach. There may be free-market policies that work in some instances and government regulatory requirements that must be applied in others. At the same time, Derr agrees with DeWitt that "selfless love in the service of stewardship of the earth" must be the principle that guides Christian action.

Congressman Vernon Ehlers writes as both a scientist and a politician. He questions the fact that DeWitt places so much emphasis on the "number one" religion, because there are other false religions, including civil religion, that do as much damage today as self-centered consumerism. Moreover, there are additional issues that DeWitt does not take up that concern Ehlers. These include the tension between doing justice to property owners and doing justice to the environment, and the tension between population growth and protecting human procreation. Ehlers applauds DeWitt's efforts to influence legislation and wants Christians to become more politically involved. Yet Ehlers stresses in conclusion that political action should be well informed, persistent, and well organized, displaying sincere appreciation for the complex responsibilities that public officials have.

The Kuyper Lecture is named in honor of the influential Dutch statesman, theologian, educator, and journalist Abraham Kuyper (1837–1920), who inspired and organized Christians for service in each of these arenas. By means of its lecture series, the Center for Public Justice wants to make the public conscious of three dynamics at work in the world today: (1) the driving influence of competing worldviews on public life, (2) the comprehensive claims of Jesus Christ on the world, and (3) the importance of the international Christian community.

In the coming decades, the Center will take the challenges of this volume to heart by promoting public-policy research and civic-education programs that encourage government and citizens to do greater justice to the environment. We hope this book will encourage many readers, from all parts of the world, to work with us toward that end.

Introduction

This lecture series is named in honor of Abraham Kuyper, and it was in the Kuyperian and Reformed tradition that I was raised. What that meant in my childhood and youth was that the whole world should be affected by the transforming power of the gospel. The world was our proper place and we were to do our part, with full energy and vigor, to repair and renew this world that we have upset and broken. It was our privilege to image and to honor God in our caring for his world, polishing it to new luster.[1]

For me this immense breadth and reach of the gospel came to fullness when, as a student at Calvin College, I read the Stone Lectures delivered by Kuyper in 1898 at Princeton University. In reading these, I was assured that all areas of life—politics, art, science, and everything else—should be transformed by the gospel. Kuyper said that Reformed Christianity brought honor to every human being "for the sake of his likeness to the Divine image," but also "to the world as a Divine Creation." As he stated, "Henceforth the curse should no longer rest upon the *world* itself, but upon that which is *sinful* in it, and instead of monastic flight *from* the world the duty is now emphasized of serving *in* the world, in every position in life."[2]

This is my heritage, and it is because of this heritage that I became a scientist. I knew science to be a God-given means for understanding God and God's world—what it is and how it works.[3] And I was also to keep theology alive and well, for it also is a means for understanding God and God's Word.[4] The two went together, hand in glove. And this was reinforced in my youth when, waiting for the service to begin, I read with pleasure the second article of the Belgic Confession from the back of the *Psalter Hymnal*. It said that we know God "first, by the creation, preservation, and government of the universe, since the universe is before our eyes like a beautiful book in which all creatures, great and small, are as letters to make us ponder the invisible things of God: his eternal power and his divinity. . . . Second, He makes Himself more clearly and fully known to us by his Holy and divine Word."

All was integrated into a seamless fabric—a Reformed world-and-life view, we called it. Everything I did had some sense of God's blessing: my backyard zoo, learning my catechism, playing my trombone, and studying science and theology. God is Author of world and Word, and all could be fruitfully studied to learn more of our Creator and God's will for our lives. It is from this integrative perspective that I became troubled by something that was emerging as a perplexing puzzle and that now has taken a front seat in many of our churches. We have, I think, been dismembering our Creator.

1

A Perplexing Puzzle
in the Context of Geo-Crisis

The Kuyper Lecture series addresses three dynamics that are at work in contemporary culture: (1) the driving influence of competing worldviews on public life, (2) the comprehensive claims of Jesus Christ on the world, and (3) the strength of the international Christian community. My topic, Responsible Stewardship of God's Creation, touches on each of these dynamics. I will begin by setting the context of my topic. I will then build toward three big questions, the answers to which have overwhelming consequences for the church and the world. Finally, I will set forth a challenge, which if taken positively, looks to a revival of the biblical truths about the Creator and creation that have served us well up until our present age.

The Pieces Do Not Fit

A perplexing puzzle lies at the heart of my topic: The claims of Jesus Christ on the world as Creator, Integrator, and Reconciler are largely neglected by those who follow

15

Jesus Christ. Instead, many Christians overlook, neglect, and in some cases even despise Christ's creative, sustaining, and reconciling works in creation. These works of Jesus Christ are expressed in the beautiful hymn of Colossians 1:15–20: "[Christ] is the image of the invisible God, the first-born over all creation. For by him all things were created . . . all things were created by him and for him. He is before all things, and in him all things hold together. . . . For God was pleased to have all his fullness dwell in him, and through him to reconcile to himself all things."

Jesus Christ is Creator, Integrator, and Reconciler, yet many who call on his name abuse, neglect, and do not give a care about creation. That irony is there for all to see. Honoring the Creator in word, they destroy God's works in deed. Praising God from whom all blessings flow, they diminish and destroy God's creatures here below. The pieces of this puzzle do not fit! One piece says, "We honor the Great Master!" The other piece says, "We despise his great masterpieces!"

Of course, Christians are not alone among the degraders of God's creative works in the world. People from a wide variety of belief and disbelief accompany them in bringing degradation to God's creation. Yet creation's degradation by those who confess the Creator while trampling, muddying, and degrading the Creator's works remains a great puzzle. Why should those who love and honor the Creator act in creation as though they despise God's masterpieces that are displayed across the great canvas of the biosphere and the heavens beyond?

The context of these two pieces of a perplexing puzzle is a global situation of serious magnitude. It is a situation in which the creating, sustaining, and reconciling work of Jesus Christ in creation is being contested through human actions that are uncreating the world, degrading its integrity, and abusing creation. The result of these actions is often labeled "the global environmental crisis."

While not having the character of crisis on a human timescale,[1] it does have such character on a geological timescale. Building for a century and more, it hardly is a sudden happening by human time; building for less than a millennium, it is a sudden event in geological time. Thus, geologically we can call a "crisis" what on the annual calendar of human events is chronic.

What is this eco-crisis? It is the crisis of one peculiar and special species having amplified its presence to such an extent that it has become a major geological force. The human presence in God's creation has reached geologically significant proportions.[2] This crisis can be summarized in seven degradations of creation.

Seven Degradations of Creation

1. Alteration of planetary energy exchange with the sun is bringing about global warming and destruction of the earth's protective ozone shield. We have been doing this by altering the concentration of greenhouse gases responsible for regulating the earth's temperature[3] and adding ozone-depleting chemicals to the stratosphere, thereby reducing the capacity of the atmosphere to filter out damaging ultraviolet-B radiation.[4] These activities are seriously threatening the earth at the fundamental planetary level.[5]

2. Land degradation is reducing available land for creatures and crops and destroying land by erosion, salinization, and desertification. In the United States we convert annually over one million acres of cropland into urban and urban-related uses, out of a cropland base of 400 million acres. In our early history, farming communities were located on our best land, and so were the feed stores, mills, and churches. As we introduced labor-saving technology, people moved to occupy these points, transforming them into villages and cities. More recently it has become economically feasible for urban people to purchase land within

commuting range of cities. Similar processes are occurring around the globe. In addition, much of the
world's agricultural land is being degraded in quality
by erosion,[6] desertification, and salinization.[7]

3. Each year, deforestation removes 100,000 square kilometers of primary forest and degrades an equal amount
by overuse. In tropical rain forests alone, an area the size
of Indiana is being deforested each year largely for non-
sustainable logging and grazing. The result is degraded
land, diminished biodiversity, and lost soil nutrients and
fertility. Of about 8.5 million square kilometers of remaining tropical primary forests (an expanse roughly
equivalent to the United States), nearly 100,000 square
kilometers are deforested each year and an additional
100,000 square kilometers degraded. With this comes
reduced sustainable production of forest products, reduced watershed recharge, increased flood flows, de-
creased drought flows, increased international disaster
relief, reduced evapotranspiration, reduced cloud formation, increased release of carbon dioxide, and extinction of species.[8]

4. Species are becoming extinct on a scale similar to the
greatest extinction episodes in the earth's geological
history. Destruction and pollution of habitats are ac-
celerating worldwide. Natural areas within commuting range of cities are being replaced with suburban
developments; wetland, estuarine, and coastal habitats are being degraded by wastes, draining, and fill-
ing, and all this is accompanied by degraded fisheries
and reduction of nutrient processing and flood peak
moderation. The result is a global crisis of plant and
animal extinctions. Similar losses are occurring in
agriculture by displacement of traditional, diverse,
local seed stocks by fewer, widely used, modern varieties. In marine areas and coastal regions, degradation
of habitats and populations of marine organisms, including those of commercial value, is also continuing.

Increasing urban and industrial development destroys productive coastal wetlands and reefs and affects the open oceans.

5. Water degradation is defiling groundwater, lakes, rivers, and oceans. Beneath people and ecosystems around the world lie great supplies of groundwater, often of high quality. This water is now being degraded by pesticides, herbicides, fertilizers, and leachates from dumps, landfills, and toxic waste sites. Beyond this, wastes are intentionally disposed through deep injection wells. In the developed world, and increasingly in the developing world, contamination of groundwater is a serious threat to human and environmental health.[9]

6. Global toxification is resulting in distribution of troublesome materials worldwide by food and product distribution and by atmospheric and oceanic circulations. Our modern society has produced some seventy thousand different kinds of chemicals, and many of these are part of the earth's global circulations. We know very little about the impact of these chemicals, but we do know that these are chemicals with which life on earth has had no prior experience. Earth's biosphere and ecosystems, its flow of energy, hydrological cycle, and biogeochemical cycles have been affected on a global scale by the products and by-products of human activity.

7. Human and cultural degradation are threatening and eliminating long-standing knowledge held by native and some Christian communities on living sustainably and cooperatively with creation as well as knowledge of garden varieties of food plants. Among the most serious degradation is the extinguishing of long-standing cultures that have learned to live sustainably on earth. These include Amish and Mennonite farmers as well as traditional small-farmer communities of the midwestern United States and numerous traditional cultures around the world.

As we become more aware of our effects on Planet Earth, we are beginning to ask questions about their consequences. A global perception of creation and its degradation is emerging. More than that, we are discovering that the common agent of creation's degradations is human action. We have known that our species is distinct in creation; now we are learning a new dimension of our distinctness: We human beings have the capacity to destroy any other creature, to degrade and destroy the various provisions in creation for renewal and rejuvenation, and even to destroy the earth itself.

What does all this mean for us as American Christians, as pastors of congregations, and as missionaries to people around the world? What does this mean for the practical witness of the international Christian community? What does this mean for public policy here and abroad? Do we have anything to say as followers of the Creator, Integrator, and Reconciler of all things to the uncreators, disintegrators, and disharmonizers of creation?

2

Religion
and the Environment

There is a little question that points to the big questions I will be treating in this paper. It is a little question with big consequences: Can we afford to save the Lord's Behemoth?

I have phrased this little question religiously. I have not asked, for example, whether we merely should save Behemoth—or the spotted owl or the snail darter or the furbish lousewort. Instead, I have asked whether we should save something made and owned by the Creator. More than that, I have asked the question in terms of economy; I have asked whether we can afford to do it. Asking the question in terms of a creature's maker and owner and in terms of its place in the economy of things makes this a religious question. And asking it religiously makes all the difference!

Let me illustrate. On November 27, 1991, several members of Congress wrote to the president of the National Academy of Sciences to request a study of several issues related to the Endangered Species Act. In response, a committee of the National Research Council produced a comprehensive study that was published in book form in 1995.[1] In it the committee reported that we are witnessing a major extinction of species at a rate comparable to the five mass extinctions in geological history, during which 14 percent to 84 percent of the genera, or families, disappeared from the fossil record.[2] The committee also reported that human activity is the primary cause of the current extinction, with most estimates suggesting that human activity has increased the extinction rate over the background rate "perhaps by orders of magnitude."

This scientific study left no option for the U.S. Congress but to strengthen the Endangered Species Act. Yet despite the considered scientific work of the National Academy and other scientists, the major bills written for ESA reauthorization would have seriously weakened the ESA. House Bill 2275, according to the Center for Marine Conservation, would have granted oil and gas and commercial fishing industries a license to kill endangered marine wildlife; allowed shrimp fishers to ignore effective conservation measures such as turtle excluder device (TED) requirements; authorized the timber industry to clear-cut ancient-forest nesting habitats of marbled murrelets; abandoned the ESA's central goal of recovering threatened and endangered species; and given foreign nations veto power over U.S. efforts to protect endangered species.[3] Scientific knowledge of extinction and endangered species was overcome by its critics mainly on economic grounds.

On January 31, 1996, the *New York Times* quoted me as saying, "The Endangered Species Act Is OUR Noah's Ark—Congress and Special Interests Are Trying to Sink It." This was followed by a news conference in which evangelicals testified that the issue of endangered species and species extinction was a religious one: "These are the Lord's creatures, and we are responsible to our Creator for their care and keeping." It received extensive national coverage on all the TV networks and all major newspapers and news services. Discussions were held about the meaning of Noah and the ark, conservatives and liberals considered Noah's example, Christians prayed, and Bible-believing legislators thought deeply about whom they ultimately answered to for the care of creation. God's command to Noah, perhaps the first endangered species act on record, had its impact on the Congress. Caring for and keeping endangered species clearly is a religious issue, and it received a religious response that superseded "business as usual." The remarkable outcome was that all further action on the ESA was delayed until after the November 1996 presidential election.

There is much to be plumbed from these events, but in sum it appears that religion, and evangelical Christians in particular, made the difference. Where scientific arguments did not prevail, religious arguments did. Yet the scientific report by the National Academy was vital, as were the contributions of scientists around the world. There is no doubt that our knowledge of the situation came from substantial scientific research on the subject. While scientific knowledge of the situation informed the evangelicals, however, it was their religiously deep response to this knowledge that got to the hearts of the American people and their legislators. Science and religion found themselves in partnership in caring for God's earth.

Here was the first clear indication that the biblical claims about God's creation were having an impact. People who were convicted by biblical teachings, believing that the Creator's care for the creatures and creation should be reflected in our own care and keeping, spoke out in behalf of God's creation. They were heard. They were respected. They "gave them religion."

While religious considerations are vital in every legislature, parliament, or royal court, it is the scientists today who are raising the ethical questions about creation's degradation. As Douglas John Hall has observed, "More than any other single segment of the general public today—certainly more than government leaders, lawyers, philosophers, and educators—more, even, than most mainline preachers, it is the scientists who are telling us that our world is in critical shape and that the human element is chiefly to blame for it."[4] Environmental scientists whose business it is to study biospheric and planetary processes and investigate environmental degradations are calling for action that will stop and reverse these degradations. By their statements and actions, they are declaring such degradations to be wrong, to be contrary to the way the world works, and they are telling us that people are chiefly to blame for them.[5]

Joint Appeal by Religion and Science for the Environment

Scientists with these ethical concerns recently have joined with religious leaders to address creation's degradation. The 1992 Joint Appeal by Religion and Science for the Environment stated,

> We humans are endowed with self-awareness, intelligence and compassion. At our best, we cherish and seek to protect all life and the treasures of the natural world. But we are now tampering with the climate. We are thinning the ozone layer and creating holes in it. We are poisoning the air, the land and the water. We are destroying the forests, grasslands and other ecosystems. We are causing the extinction of species at a pace not seen since the end of the age of the dinosaurs. . . . The magnitude of this crisis means that it cannot be resolved unless many nations work together. We must now join forces to that end.

For those who invest their lives in its study, the world teaches sufficiently about its own integrity that deviations are clearly identifiable. The integrity of creation is everywhere apparent, and it is this integrity that in fact makes science possible.

Out of the Joint Appeal came the formation in 1993 of the National Religious Partnership for the Environment—a partnership that includes the Coalition on the Environment and Jewish Life, the U.S. Catholic Conference, the National Council of Churches, and the Evangelical Environmental Network (EEN). It was the last of these that made the announcement that the Endangered Species Act is *our* Noah's ark, an announcement that had the full concurrence of the others. Others have also recognized the tremendous significance of religion in our time, and in particular the significance of the church. Respect for religion is also growing among secular ethicists.[6]

Religious responses and religions are not necessarily effective on environmental issues, however. Chinese-American geographer Yi-Fu Tuan, for example, in assessing Eastern cultures, found "that, despite their different religious traditions, their *practices* were every bit as destructive of their environments as in the West."[7] The effectiveness of a religion's response to a particular situation depends upon what Huston Smith refers to as its quality.[8] Religions have not always achieved wholeness, even though they may be strong shapers of society and environment. The importance of religion is evidenced in the claim of Lynn White Jr. that due to its power, the Judeo-Christian religion has been the root of our ecological crisis.

It is not a foregone conclusion, then, that religions will help sustain creation in its integrity. Instead, they may be captured "by things that matter least," and may even become accomplices in the earth's destruction. Beyond this, Huston Smith informs us, religions may also yield to forces they attempt to challenge and constrain, thus lessening their quality and even their relevance. Religion of high quality restores the ability of people and society to master themselves, since it uncovers and revitalizes beliefs that bring respect for God and creation.

The quality of evangelical Christianity is weakened by the fact that it operates within the context of another religion. This other religion organizes all (or most) of life around its central aspirations and contends with evangelical Christianity for control over all areas of life, including politics, education, science, technology, and the media. In the spirit of our Western inventiveness, we have invented not only new things and new processes but also a new religion. We have come to believe in our day that we know enough to create in a mere geological moment systems superior to the self-sustaining ecological systems that have stood the test of geological time, and so we also have invented a new religion in a historical moment to replace and contend with the Judeo-Christian religion that has stood the test of time. This new religion has become our number one religion, but it is failing us. It is shortchanging us and leading to ecological and spiritual bankruptcy.

Something I heard from a dear Christian friend several years ago helped me unmask our new religion. I was talking with him about how things were going. It was Christmas and we were celebrating the birthday of Jesus. In the middle of talking about our goals in life, he commented, "Well, we have to look out for *number one*." Quite naively, I thought by his reference to number one he was talking about God, but his next few sentences cleared away that idea. He was talking not about God but about himself! We have to look out for ourselves! *We* are number one!

This expression has since become widely used. It has been incorporated into what we believe about life and the world, and it is a belief we sometimes find ourselves and others defending religiously. Because its principal doctrine is "Look out for number one," we can call it our "number one" belief. But while the expression is new, the idea is not. It is present, for example, in *The Wealth of Nations*, the work of a famous Scottish professor of moral philosophy. While its author, Adam Smith, wrote about "the natural effort of every individual to better his own condition" as its cornerstone, others came out and called it human greed and avarice. People did not feel fully comfortable with the basic idea behind this belief. The economist John Maynard Keynes, for example, envisioned it operating for a while, after which we could return to an economy driven not by greed but by Christ's teachings and traditional virtue.[9] A more recent practitioner of this "number one" economy, Charles Schultze, advocated remaining in the "tunnel of economic necessity" because it reduces the need for virtue.

> Market arrangements not only minimize the need for coercion as a means of social organization, they also reduce the need for compassion, patriotism, brotherly love, and cultural solidarity as motivating forces behind social improvement. Learning how to harness the "base" motive of material self-interest to promote the common good was perhaps *the* most important social invention mankind has yet made. Turning silk into a silk purse is no great shakes, but converting a sow's ear into a silk purse does indeed partake of the miraculous. . . . There is indeed a role for "preaching" as a means of creating a political and cultural situation in which consensus can be reached on social intervention. Cleaning up the environment will only be achieved as environmental quality takes a higher place in the esteem of most citizens. But when it comes to the specifics of getting the job done, preaching, indignation, and villain identification get in the way of results.[10]

This in a nutshell is a description of our "number one" belief, our "number one" economy. Its proof text is 33:6 .ttaM: "Seek first yourself, and the kingdom will be added unto you." 33:6 .ttaM puts backward Matthew chapter 6, verse 33: "Seek ye first the kingdom of God, and his righteousness; and all these things shall be added unto you" (KJV)[11]. Setting this backward verse as a cornerstone in the human economy performs a remarkable flip on how we think about greed. It takes greed, long recognized as a vice, and declares it to be a virtue. The consequence is the miraculous "converting a sow's ear into a silk purse."

This backward thinking—now being adopted across the globe—thrives on an invention that makes it no longer necessary to refer to Number One—the Maker of the heavens and the earth. Its only reference is number one—ourselves! In our "number one" economy, we have figured out a way to avoid any need for the divine Economist through whom the whole world was made, is sustained, and is reconciled (see Col. 1:15–20). This "number one" economy is an important piece of the puzzle for understanding responsible steward-ship of God's creation.

Some would say that we are speaking here of two differ-ent things: the "number one" economy of the world (in which number one refers to us) and the Number One econ-omy of the church (in which Number One refers to God). The first belongs to the public sphere; the second belongs to the private sphere. Abraham Kuyper warned us, however, against adopting "the belief, which took hold in the West following the French Revolution, that religion can be con-fined to a private sphere and that other areas of life can be treated as religiously neutral (or nonreligious) and managed through a secular consensus."[12] Kuyper recognized that the Enlightenment notion of a division of life into sacred and secular compartments functions as a religious driving force that when successful, takes control of human hearts and minds. The modern era, Kuyper maintained, simply cannot be understood without grasping the significance of the driv-ing force of competing religions—religions both ancient and modern, theistic and secularist.

By unmasking our "number one" religion, we discover an approach to life that is functionally religious, whether or not its participants call themselves religious. Our "number one" economy has become our way of doing business, our way of doing our jobs. Sometimes we even find ourselves defend-ing this way of living with religious zeal, as if our salvation depended on it.

When we take the mask off our "number one" religion, we see that it almost always means that it will have us make im-pious use of creation, simply because we are number one. This has its consequences, including a kind of judgment by creation itself. Joseph Sittler, the Lutheran theologian, wrote in this regard that

> if the creation, including our fellow creatures, is impiously used apart from a gracious primeval joy in it, the very richness of the creation becomes a judgment. . . . When things are not used in ways determined by joy in the things themselves, this violated potentiality of joy . . . withdraws and leaves us, not perhaps with immediate positive damnations but with something much worse—the wan, ghastly, negative damnations of use without joy, stuff without grace, a busy, fabricating world with the shine gone off.[13]

"With the shine gone off"—no more luster! At the heart of our "number one" religion is an economy of use only. It is use that must be fueled and sustained by dissatisfaction with what we already have. It fosters an economy of discontent that festers at the heart of our restless society, putting forth survival of the economically fittest as an expected outcome. Besides that, our "number one" economy reduces the creation to resources and people to consumers. It has no need for God in making decisions and has no authority except the free self. It determines value by price, and price by willingness to pay. It is largely unable to conserve ecosystems and landscapes. It denies intrinsic good, and it presumes to operate independently from the larger economy of creation.[14]

We have put one of God's many gifts for the distribution of goods, the market, into a position through which we try to subsume God's entire economy for creation. We have placed it in a position of such authority that everything in creation is converted to the status of resources to be bought, sold, and used in what is called the marketplace. As the only authority in the "number one" religion, the market is respected and revered and defended religiously. Thus, we have extended our use of this creature, the market, beyond its role as a means for responsible stewardship in behalf of the Creator to a replacement of the need for a Creator altogether. Our "number one" religion has led us to worship this creature rather than the Creator. Even the name of this creature is held in reverence. We defend it and its name at all costs. God's gift has become the people's god. It is the only god-meriting global worth-ship. All other religions accommodate the "number one" religion, since the "number one" religion tells everybody what must be done and what ought to be. The market, rather than serving as a means for stewardship, has been elevated to the arbiter of our personal and global ethics, with the result that human beings are divested of their role of stewards of creation and are seen as mere consumers of creation.

God's Economy as *Oikonomia* of the Cosmos

Evangelical Christianity claims that God owns all things and that God is our model for what we must do. Moreover, it defines our relationship to the world as one of stewardship, or caring for something (the creation) on behalf of another (God). As "all the mighty acts of God in creation are mutually a work of the Triune God—Father, Son, and Holy Spirit,"[15] so our stewardship is in behalf of the Triune God. Louis Berkhof describes it thus: "All things are at once out of the Father, through the Son, and in the Holy Spirit."[16] So in our stewardship we image the work of the Triune God in creation, responding in worship to Father, Son, and Holy Spirit.

The "number one" economy stands in stark contrast to God's economy as the latter is described in the Scriptures and as it operates in creation. Matthew 6, an important chapter on God's economy, is illustrative, including as its central premise taught by Jesus, "Seek ye first the kingdom of God." Psalm 104 is another, including its description of God giving the landscape rain in its season and providing food for the lions and other wild creatures.

Webster's dictionary is particularly helpful here. According to it, one meaning of *economy* is "God's plan or system for government of the world."[17] This meaning of *economy* has been recognized for centuries. One example comes from Carl Linnaeus (Carolus von Linné), the Swedish taxonomist who in 1791 wrote in his *Oeconomy of Nature,* "By the Oeconomy of Nature we understand the all-wise disposition of the Creator in relation to natural things, by which they are fitted to produce general ends, and reciprocal uses. All things contained in the compass of the universe declare, as it were, with one accord the infinite wisdom of the Creator."[18] Theologians of Linnaeus's time, not having the word *ecology,* made the word *oikonomia* interchangeable with God's *dispensations,* so that by the seventeenth century *oeconomy* was frequently employed to refer to the divine government of the natural world. "God's economy was His extraordinary talent for matching means with ends, for so managing the cosmos that each constituent part performed its work with stunning efficiency."[19]

Well before Linnaeus, there was the economy of Eden referred to in Genesis 2:15. A literal reading of this passage describes the human economy in Eden thus: "And Jehovah God taketh the man and causeth him to rest in the garden of Eden, to serve it and to keep it."[20] Commenting on this passage in 1554, John Calvin wrote,

> The custody of the garden was given in charge to Adam, to show that we possess the things which God has committed to our hands, on the condition, that being content with the frugal and moderate use of them, we should take care of what shall remain. Let him who possesses a field, so partake of its yearly fruits, that he may not suffer the ground to be injured by his negligence; but let him endeavour to hand it down to posterity as he received it, or even better cultivated. Let him so feed on its fruits, that he neither dissipates it by luxury, nor permits [it] to be marred or ruined by neglect. Moreover, that this economy, and this diligence, with respect to those good things which God has given us to enjoy, may

flourish among us; let everyone regard himself as the steward of God in all things which he possesses. Then he will neither conduct himself dissolutely, nor corrupt by abuse those things which God requires to be preserved.[21]

By the phrase "this economy" Calvin is referring to:

1. our contentment with the frugal and moderate use of the things God has committed to our hands,
2. our care for what we do not ourselves use,
3. our taking of the fruits of a field without letting the ground suffer through our negligence,
4. our handing down our land to posterity as good as, or better cultivated than, we received it,
5. our feeding on its fruits in a manner that neither dissipates it by luxury nor permits it to be marred or ruined by neglect.

By the term *custody*[22] Calvin interpreted dominion to mean a responsible care and keeping that does not neglect, injure, abuse, degrade, dissipate, corrupt, mar, or ruin the earth.[23] God's economy, "God's plan or system for government of the world," is always the context and framework within which the human economy works.

We learn from Genesis 2:15 that the biblical idea of economy is serving and keeping creation, not oppressing it through domination. Jesus Christ, our model, reinforces this. The image of God (2 Cor. 4:4; Col. 1:15) takes "the very nature of a servant" (Phil. 2:6–7). Christians, in working out their economy within God's economy for creation, follow the Creator-Servant, the Second Adam, joining Christ in reconciling all things to God, undoing the damage of the first Adam, and doing what Adam was supposed to do.[24] More than that, they cultivate society as part of God's creation, seeking and preserving truth,[25] building civilized societies, and establishing the church of Christ on earth.[26] They thus demonstrate that an economy that harmonizes with God's economy represents "a truly biblical Christianity" and "has a real answer to the ecological crisis."[27]

It is clear from Calvin's commentary that the question is not whether there should be a human economy but rather which of the two economies—the human economy or God's economy—should be a subset of the other. Thus, while we cannot abandon the idea of a human economy, we must be sure to put our economy within the context of God's economy. Whatever kind of economy we invent for ourselves, our economy must always be part and parcel of God's economy.

Stewardship Defines the Relationship between the Two Economies

The relationship between human economies and the economy of God's wider creation is circumscribed by the word *stewardship*. *Economy*, whether applied to the workings of cosmos or village, comes from the Greek word *oikonomia*, meaning "management of the household." The household can mean the cosmos as the household of all life on earth or the household of house and family members. *Ecology* (study of the household), *economics* (management of the household), and *ecumenical* (in Greek, *oikoumene*— the inhabited world) all share the root word *oikos*, meaning "house." Our human household, then, is part of the larger household of life, which in turn is part of the household of all God's creation. Our human relationship within and among these households is described by *oikonomia*, or stewardship.

Stewardship is our use and caring for the household on behalf of the Creator, whose stewards we are. Thus, our economy is necessarily part of God's economy.[28] Moreover, our human economy must be justly aligned within God's economy. If we take the word *harmony* not in its romantic but in its technical sense, as in music theory, the word *har-*

mony can be used as a metaphor for how the human economy should relate to God's economy.[29]

The Creator of all things, whose divine economy is the envelope in which human economies must operate, is the only one who merits worship. This God whom we worship requires us to act justly and to love mercy and to walk humbly with our God. This is the core of stewardship of creation and the Creator's gifts. Our worship of God and our just stewardship of God's world go hand in hand. Worship of the market and consumerism are compatible with each other, but worship of God and consumerism are not.

Noah provides an instructive model of a person who is obedient to the Creator and who faithfully learns to understand the needs of creatures under threat of extinction, supplying what is needed to perpetuate their lineages. Noah engages in freedom *for* responsibility. Free to choose between life and death, he chooses life (cf. Deut. 30:11–20). This stands in sharp contrast to people in the wider society who have chosen to exercise freedom *from* responsibility even though it is within their reach to walk in the ways of their Creator and to observe the laws of creation.

Ignorance of God's economy is no excuse. That is why we must make a serious effort to know the workings of God's economy in creation and make sure that we do not obscure our and others' understanding of these workings. Not knowing, for example, that floodplains are God's provision for handling floodwaters and that barrier islands are for breaking the force of the sea is no excuse for calling catastrophic destruction of houses and hotels on floodplains and barrier islands "acts of God."

3

The Three Big Questions

Can Christianity provide an effective response to the need for human care of creation? It depends on the answers we give to the big questions. These are the three big questions we must ask in this global crisis if we confess that Christ is the Creator and the Great Integrator and Reconciler of all things.

1. Is Jesus Christ Lord of Creation?
2. Is creation a lost cause?
3. Whom are we following when we follow Jesus Christ?

Question 1: Is Jesus Christ Lord of Creation?

Through Jesus Christ, God created the world, holds everything together, and reconciles all things (Col. 1:15–20). Followers of Jesus Christ have known this remarkable teaching of Colossians from the beginning. The depth and significance of this passage are there for all to see.

[Christ] is the image of the invisible God, the firstborn over all creation. For by him all things were created: things in heaven and on earth, visible and invisible, . . . all things were created by him and for him. He is before all things, and in him all things hold together. And he is the head of the body, the church; he is the beginning and the firstborn from among the dead, so that in everything he might have the supremacy. For God was pleased to have all his fullness dwell in him, and through him to reconcile to himself all things, whether things on earth or things in heaven, by making peace through his blood, shed on the cross.

We observe three things about this passage of Scripture. First, Christ is the Creator of all things *(ta panta)*, the Author of all things. Not only did he create all things but all things were created for him. Second, in Christ all things *(ta panta)* hold together. Everything derives its integrity from Christ, and without him things would fall apart. Christ is Sustainer and Integrator of all things. Third, God reconciles all things *(ta panta)* to himself through Christ by making peace through his blood, shed on the cross. So Christ is Reconciler and Harmonizer of all things.

The Scriptures make it clear that the claims of Jesus Christ on the world are comprehensive. It is the claim made by the one who made the world, holds the world together, and reconciles the world. The comprehensive claims of Jesus Christ on the world derive from his being its Author, Integrator, and Harmonizer.

Second Corinthians 5:15 takes us further in our understanding of the consequences of this claim of Jesus Christ on the world. There it states that "he died for all, that those who live should no longer live for themselves but for him who died for them and was raised again." This has a remarkable consequence: namely, that living in Christ, we no longer regard Christ or anyone else from a worldly point of view. In Christ we are a new creation and live as his ambassadors, committed to the message of reconciliation (2 Cor. 5:16–20).

What, then, does it mean to be Christ's ambassadors? It is inconceivable that those who are in Christ and who themselves have been made new creatures should find themselves in opposition to Christ's work of creation, integration, and reconciliation. Can we honor our Creator without respecting his creation? Can we honor our Creator and despise his magnificent works? Can we thank God for loving the world and not care about it?

My purpose in all of this is to help enlarge our understanding of the lordship of Jesus Christ so we appreciate that Jesus Christ is Creator, Sustainer, and Reconciler. My conclusion is this: Jesus Christ *is* Beautiful Savior, but more than that, Jesus Christ *is* King of Creation. Have we dismembered our Creator into Savior on the one hand and Maker of the material world on the other? Have we separated the spiritual from the material? Are we treating the great gallery of our Creator in a way that expresses our love and respect for our Triune God?

Question 2: Is Creation a Lost Cause?

Because of the degradation human beings have brought across the face of the earth, we are *tempted* to ask whether creation is a lost cause. I use the word *tempted* here deliberately, for it is in fact a temptation to consider giving up on creation altogether, and many have done so. The temptation that entices us is this: that we should live as best we can for now and pin our hopes on the life to come. Yielding to this temptation may bring us to a position where we do not care about the material world at all. We would profess not only that matter does not matter but that matter in the material world is ugly, perhaps even evil.

It is devastating to ourselves and to God's world when we yield to this temptation. We should be put on guard by the warning of Revelation 11:18, where in the last judgment, following the sounding of the seventh trumpet, the proclamation is made that "the time has come for . . . destroying those who destroy the earth." On the positive side, there is God's love for the world, God's love for the cosmos, a love so great that God takes on material flesh. There is also the powerful taking on again of the flesh in Christ's resurrection. So we must not yield to this temptation, since to God incarnate, matter matters!

Abraham Kuyper has a contribution to make here. It comes by way of an exposition he gave in 1903 of John 3:16 ("For God so loved the world") over half a century before our recognition of a crisis, even before any popular understanding of ecology or "the environment."

> So God loved the world, that He gave it His Only-begotten Son. . . . God loves *the world*. Of course not in its sinful strivings and unholy motions. . . . But God loves the world for the sake of its origin; because God has thought it out; because God has created it; because God has *maintained* it and *maintains* it to this day. We have not made the world, and thus in our sin we have not maltreated an art product of our own. No, the world was the contrivance, the work and the creation *of the Lord our God.* It was and is His world, which belonged to Him, which He had created for His glory, and for which we with that world were by Him appointed. It did not belong to us, but to Him. It was His. And it is *His* divine world that we have spoiled and corrupted.
>
> And herein roots the love of God, that He will repair and renew this world, His own creation, His own work of wisdom, His own work of art, which we have upset and broken, and polish it again to new luster. And it *shall* come to this. God's plan does not miscarry, and with divine certainty He carries out the counsel of His thoughts. Once that world in a new earth and a new heaven shall stand before God in full glory.
>
> But the children of men meanwhile can fall out of that world. If they will not cease to corrupt His world, God can declare them unworthy of having any longer part in that world, and as once He banished them *from Paradise,* so at the last judgment He will banish them *from this world.* . . .

And therefore whoever would be saved with that world, as God loves it, let him accept the Son, Whom God has given to that world in order to save the world. Let him not continue standing afar off, let him not hesitate.[1]

Clearly, then, creation is not a lost cause. God expresses his eternal love for the world by giving us his Son. Reflecting on this gift in the context of what we have learned from Colossians 1:15–20, we are struck by the fact that God's gift to the world pre-exists the world. God's gift to the world, Jesus Christ, is before all things. Yet he comes in the flesh. The Creator takes on created matter as part of himself. The material creation matters to God.

Beyond the incarnation is the resurrection of Jesus Christ. In it we have the conclusive answer to the question of whether creation is a lost cause. It is through the resurrection in particular that creation is vindicated. Evangelical ethicist Oliver O'Donovan puts it this way:

> We are driven to concentrate on the resurrection as our starting-point because it tells us of God's vindication of his creation, and so of our created life . . . .
>
> The meaning of the resurrection, as Saint Paul presents it, is that it is God's final and decisive word on the life of his creature, Adam. . . . It might have been possible, we could say, before Christ rose from the dead, for someone to wonder whether creation was a lost cause. If the creature consistently acted to uncreate itself, and with itself to uncreate the rest of creation, did this not mean that God's handiwork was flawed beyond hope of repair? It might have been possible before Christ rose from the dead to answer in good faith, Yes. Before God raised Jesus from the dead, the hope that we call "gnostic," the hope for redemption *from* creation rather than for the redemption *of* creation, might have appeared to be the only possible hope. "But in fact Christ has been raised from the dead . . ." (15:20). That fact rules out those other possibilities, for in the second Adam the first is rescued. The deviance of his will, its fateful lean-

ing towards death, has not been allowed to uncreate what
God created.[2]

The resurrection of Jesus Christ means that the creation
is not a lost cause. Creation is affirmed by its Creator.

We continually confront in the church the devilish temptation to reduce the Lord of Creation to one who merely
saves. Under the continuing influence of the Gnosticism
that infected the early church, some have reduced God to
the one who saves us out of creation. This separation of Savior from Creator goes so far that belief in the Creator is reduced to empty words. As art critics might somehow find it
acceptable to trample Rembrandt's paintings while honoring Rembrandt's name, some people praise the Creator
while trampling on his creation. Regrettably, some God-
praising people have comfortably neglected creation's evangelical testimony[3] and even assist in bringing about creation's degradation.

Saving people "out of creation" is not a biblical idea, of
course. Instead, it is rooted in the Platonic notion that
physical nature is a source of ignorance and evil and a
snare to the soul. When joined with the idea of human
transcendence, this resulted in a theology that "laid most
stress on the salvation of the *soul,* and which tended to dis-
miss as insignificant the body and the creation of which it
was a part."[4]

Question 3: Whom Are We Following
When We Follow Jesus Christ?

We sometimes sing, "Christ shall have dominion, over
land and sea." Jesus Christ, the Lord of Creation, is our
model for dominion, but what is that model? The apostle
Paul puts it this way in his letter to the Philippians, 2:5–8:
"Your attitude should be the same as that of Christ Jesus,
who, being in very nature God, did not consider equality

with God something to be grasped, but made himself nothing, taking the very nature of a servant, being made in human likeness. And being found in appearance as a man, he humbled himself and became obedient to death—even death on a cross!"

Jesus Christ, the Son of God, did not consider equality with God something to be grasped. Even more so we, the followers of Jesus Christ, should not view ourselves equal to God or Jesus Christ. Far from being equal with God, we must confess our total dependence upon God in every aspect of our lives and vocations. The example of Jesus Christ, our model of dominion, helps interpret for us the dominion material in Genesis 1:26–28. Taken in the context of the example of Jesus Christ, this passage helps us understand our responsibility toward the Lord's creation. The passage reads as follows (KJV, emphasis mine):

> And God said, Let us make man in our *image,* after our likeness: and let them have *dominion* over the fish of the sea, and over the fowl of the air, and over the cattle, and over all the earth, and over every creeping thing that creepeth upon the earth.
>
> So God created man in his own *image,* in the *image* of God created he him; male and female created he them.
>
> And God blessed them, and God said unto them, Be fruitful, and multiply, and replenish the earth, and *subdue* it: and have *dominion* over the fish of the sea, and over the fowl of the air, and over every living thing that moveth upon the earth.

In this passage, the Hebrew word *radah* is translated "have dominion" (KJV) or "rule" (NIV). An even more forceful word is "subdue," a translation of the Hebrew *kabash.* Without the example of Jesus Christ, one might conclude that this passage suggests "anything goes."[5] However, Jesus Christ brings us to see this dominion as service rather than as a license for ungodly behavior.

It also is clear from the requirements for kings in Deuteronomy 17:18–20 that those to whom God gives dominion must fully reflect God's will in their rule. They must reflect God in the way they relate to their subjects—mirroring, representing, reflecting, and imaging God's will and God's relationship to creation.[6] Thus, God's proclamation through Ezekiel, "Woe to the shepherds of Israel who only take care of themselves! Should not shepherds take care of the flock? You eat the curds . . . but you do not take care of the flock. . . . You have ruled [*radah*] them harshly and brutally" (Ezek. 34:2–4).

The Lord shows by divine example what should be done: "I myself will search for my sheep and look after them. . . . I will pasture them on the mountains of Israel, in the ravines and in all the settlements in the land. . . . I myself will tend my sheep and have them lie down, declares the Sovereign LORD. I will search for the lost and bring back the strays. I will bind up the injured and strengthen the weak, but the sleek and the strong I will destroy. I will shepherd the flock with justice" (Ezek. 34:11, 13, 15–16).

Lest people take the mandate to subdue the earth as a license to serve self rather than God and creation, God judges between those who use creation with care and those who abuse it. "Is it not enough for you to feed on the good pasture? Must you also trample the rest of your pasture with your feet? Is it not enough for you to drink clear water? Must you also muddy the rest with your feet? Must my flock feed on what you have trampled and drink what you have muddied with your feet?" (Ezek. 34:18–19).

What, then, is dominion in biblical and Christian terms? What does it mean to subdue the earth? It is serving God and creation. It is reflecting God's love for the world, God's law for creation, and God's justice for the land and creatures. Without this responsibility we might have little reason other than pragmatic and utilitarian ones for keeping creation. We might work to save and nurture only what is useful or attractive to us. But as Noah in his obedience to God's will worked to perpetuate the fruitfulness of endangered species, so we must think not only of the animals that Noah kept but of the other creatures as well.

Two Kinds of Dominion

Dominion as domination is forbidden. Dominion as stewardship is required as a God-given responsibility for all people. Human dominion, however, is exercised across a broad spectrum, one end of which is dominion exercised in behalf of self and the other dominion in behalf of creation. Dominion at the first extreme can be called domination; dominion at the other extreme can be called stewardship. More specifically, in relation to creation, domination is service in behalf of self at the expense of creation; stewardship is service to creation in behalf of the Creator. Thus, we can distinguish between two kinds of dominion: domination and stewardship.[7]

Much of Genesis 1–11 is addressed to the wrongness of domination. The scriptural view is that seeking first one's own selfish interests at the expense of creation and its creatures is sinful domination worthy of punishment, even death. Thus, Adam and Eve's pressing into service the forbidden fruit to know good and evil is domination; its consequence is their own death and degradation of the ground. The murdersome domination of Cain over Abel results in Cain being cursed by God, made restless, driven from the land, and the ground no longer yielding crops for him. The corrupting of creation by human society brings with it the deluvian destruction.[8] The rebellious event at Babel, where people "undertook a united and godless effort" to make for themselves "by a titanic human enterprise," a worldwide reputation and renown through which they "would dominate God's creation," results in their scattering across the earth.[9] In scriptural language, domination, defined as seeking first ourselves at the expense of creation, is "missing the mark," it is failing to meet the Creator's expectations for us, it is sin.

The scriptural view is that seeking first to do the will of God with respect to creation is the right exercise of dominion. We have seen from Genesis 1–11 what stewardship is not; from Genesis 2:15 we can learn something of what it is. Here we learn that Adam and his descendants are expected to *serve* and *keep* the garden. The word keep is a translation of the Hebrew word *shamar,* which is also used in the Aaronic blessing given in Numbers 6:24 (emphasis mine): "The LORD bless you and *keep* you," a blessing very widely used in Jewish and Christian congregations to this day.[10]

When we invoke God's blessings to keep the assembled people, we are not praying merely that God would keep them in a kind of preserved, inactive, uninteresting state, like one might keep a museum piece, a preserved specimen, or pickles in a jar, but rather that God would keep them in all their vitality, with all their energy and beauty. This keeping is one that would nurture all life-sustaining and life-fulfilling relationships—with family, spouse, and children, with neighbors and friends, with the land that sustains human life and the living creatures, with the air and water, and with God. *Shamar* is an extremely rich word with a deeply penetrating meaning that evokes a loving, caring, sustaining keeping.

This is also the thrust of Genesis 2:15. When we act on God's will and charge to keep the garden, we make sure that the creatures under our care are maintained with all their proper connections with members of the same species, with the many other species with which they interact, and with the soil and air and water upon which they depend. The rich and full *keeping* that we invoke with the Aaronic blessing is the kind of rich and full keeping that we should bring to the garden of God, to God's creatures, and to all creation. As God keeps those who invoke divine keeping, so those whom God keeps keep creation. Human beings should be engaged in the care and keeping of creation, with all the richness and fullness this implies.[11]

In addition to recognizing the fullness of the meaning of *shamar* in Genesis 2:15, it is also helpful to our understanding of stewardship to attend to the preceding word, *abad.* In *Young's Literal Translation of the Holy Bible* the passage is rendered, "And Jehovah God taketh the man and causeth him to rest in the garden of Eden, to *serve* it and to keep it." Here *serve* is a translation of the word *abad.* For those who have heard this translated "cultivate," "till," or "dress," this may come as a surprise. The word *abad* is also used in another famous passage: "Choose for yourselves this day whom you will serve. . . . But as for me and my household, we will serve the LORD" (Josh. 24:15). A search of the use of *abad* will show that it is translated "serve," as in Joshua 24:15, except when it is applied to agriculture. No matter how one deals with the proper translation of the text, however, whether as translated with agricultural language or literally as "serve," the idea of service comes through clearly.

While serving the garden or serving creation might sound peculiar to modern ears, we might consider what the Garden of Eden consisted of. The Scriptures say that it was planted by God (Gen. 2:8), which makes us wonder how God plants things in creation. In God's garden, hoe, shovel, and plow might simply have been out of place, especially if it was more like the gardens of some tropical peoples, where interplanting and high diversity are the rule. Perhaps it was a tropical garden not amenable to turning over the ground but still open to service. Whatever the case, to understand the meaning of stewardship, one must ponder the meaning of service.

The biblical expectation that human beings will "serve the garden" means that our dressing, tilling, and tending are done as acts of service. With the prefix *con,* this can be applied to indicate "service with," as with the word con-serve. We may take this to mean that as the garden serves us, so we should provide reciprocal service. The biblical expectation,

I believe, is con-service. As stewards of a creation, whether our stewardship is over a small garden or the whole biosphere, we are expected to be about the business of con-servancy. Conversely, when human beings fail to serve and instead abuse the garden or creation, they should expect payment back in kind. Intended abuse of creation can have severe consequences, as we know full well. Unintended abuse of the garden can also have severe consequences, but if we are committed to tending the garden, the repercussions of such abuse can be part of our stewardship education. The key to proper service always is to consider our service as Christ's service. Our service should reflect God's love for the world.

When dominion is taken as license to do whatever meets one's self-interest, it is a misappropriation of the image of God, and a failure to follow the example of Jesus Christ. Responsible appropriation of this image is to seek first not self but the kingdom of God. To image God is to image God's love and law. To be made in the image of God is to be endowed with dignified responsibility to reflect God's goodness, righteousness, and holiness. It is to use our intellectual powers, natural affections, and moral freedom to reflect the wisdom, love, and justice of God. It is to commune prayerfully with God through Jesus Christ and to anticipate the ultimate fulfillment of all things in the way we live our lives.

We conclude from all this that human beings are distinct with respect to other species in their exercise of dominion over creation. To the extent that being made in the image of God confers upon human beings what is distinctive with respect to other species, the exercise of dominion is part of the consequence of humans being made in the image of God. Failure to seek God's purposes in creation leads to a perverted and sinful dominion, a domination whose goal is serving self rather than the Creator or creation. The proper exercise of dominion by human beings who seek truly and fully to mirror God's wisdom, love, and justice is stewardship. So human beings should make every attempt to overcome the forces that would compel them to dominate creation, and by diligently seeking creation's integrity, vigorously and prayerfully pursue a life of stewardship with God's kingdom as its goal.

Resorting to the idea of stewardship does not eliminate the symptoms, but rather addresses problems at their roots. People can mislead themselves and each other—through selfish intent, ignorance, or denial—by confusing symptoms with the underlying causes of environmental problems. A sound understanding of the idea of stewardship will help us appreciate the difference.

The biblical imperative, then, is for stewardship in behalf of God's creation no matter what its condition. Christian environmental stewardship is not crisis management but a way of life. God's call to serve and keep the garden is our calling no matter whether it is our vegetable garden or the whole of creation, and no matter if it is being degraded, staying the same, or improving. Caring for creation is much like caring for families—in sickness or health, in riches or poverty, in crisis or harmony. And this caring must be done wisely. Caretakers must be ever aware, alert, and vigilant in a sinful world, alert even to our own sinfulness (Rom. 3:23). We must face squarely the human predicament (Rom. 7:7–25) and also be ready to give a response in word and deed for the faith that is within us (1 Peter 3:15). We need not have all the data, but we must be dedicated to imaging God's love for the world in our lives and landscapes. Responsible stewardship is not an option but a continuing privilege and responsibility.[12]

4

Can We Afford to Keep the Lord's Behemoth?

If responsible stewardship of God's creation is a continuing privilege and responsibility, then how do we address the claims that saving species or taking care of the environment just costs too much money? In the context of the big questions and our answers to them, how do we respond to a world in which people tell us we simply cannot afford to care for God's creatures? Can we afford to keep Behemoth? There is no better response than to come into the presence of God in his conversation with Job. Perhaps we should also come into the presence of Behemoth in its native habitat in some of creation's wetland ecosystems. But who is Behemoth?

There are some creatures toward which human beings have some natural affinity: soft furry ones with big round eyes, flashy feathery ones with docile beaks, colorful scaly ones with whiskery fins. But those with flabby flesh and oozing pores, with bulging throats and acrid warts, are "horses of a different color." Thus, there is likely to be an asymmetry in the affection we bestow across the creaturely spectrum. With such imbalance may come differential treatment, varying respect, and diverging care and keeping. Koalas and cockatiels are likely to be loved, cuddled, and cared for more than toads and hippopotami.

Beautiful in the Eyes of God

It is perhaps for this reason that God finds it necessary in the presence of Job to praise the unlovely, uncuddled, and little-cared-for kinds. God's psalm to the hippopotamus (Job 40:15–24) is a wondrous example. From it we can learn a great deal about respecting creatures we may not find attractive. Through the appreciative eyes of God we may see what to human eyes may seem ugly.

> [Behold] behemoth, which I made along with you
> and which feeds on grass like an ox.

This great bestial creature is my creation; I, your Creator, also made this great beast!

> What strength he has in his loins,
> what power in the muscles of his belly!

Truly magnificent my hippo is; my creature is strong and powerful!

> His tail sways like a cedar;
> the sinews of his [stones] are close-knit.

People may not want to admire the perfection and wonder of this creature's reproductive organs, but I, their Creator, am proud of this wonderful provision. Could there be a better way to provide for the procreation of such a massive creature?

> His bones are tubes of bronze,
> his limbs like rods of iron.

What architecture there is in this, my beast—strength of limb and bone, for sure! And so marvelously wrought that this creature's mass is mightily upheld.

> He ranks first among the works of God.

Not that people are not important, of course; I put it this way so they may know the importance of this creature in my sight. I rate this hippo number one—I hope you do, too!

> [Only I, the Maker of this beast,] can approach him with his sword.

As its Author, I have the authority to kill this wonderful creature, but only I. I say this not to tell you I am going to destroy it, for I am not; I say this to tell you, "Keep your hands off my hippo! Put down your arrow and gun." Don't you dare try to demonstrate your power by destroying what I through my power have made. Real power, after all, is shown by the Maker, not by the destroyer. This work of art belongs to the Creator alone, the one who has the right to create and destroy.

> The hills bring him their produce,
> and all the wild animals play nearby.

This creature is not sprawling over a concrete slab in one of your zoos. It is in its proper habitat, deriving food from it and relating to all the wild animals there, if only because they too bring God praise.

> Under the lotus plants he lies,
> hidden among the reeds in the marsh.

Its wetland habitat, while not very well suited for people, is nicely suited for the hippo, and the hippo is nicely suited for its aquatic habitat. It is fully integrated into the dynamic fabric of its watery, lush world.

> The lotuses conceal him in their shadow;
> the poplars by the stream surround him.

Integrated into its habitat, even this giant creature is inconspicuous. It fits into its habitat in a marvelous beastly harmony that sees the wetland vegetation embrace it with its spreading boughs and concealing leaves.

When the river rages, he is not alarmed;
>   he is secure, though the Jordan should surge against his
>   mouth.

Its integration into its environment goes beyond physical arrangements of plants and animals to include its own psyche; this beast is at home in the wetland and the river torrent.

>   Can anyone capture him by the eyes,
>   or trap him and pierce his nose?

Not only is this creature one that might not respond to your beckoning, it is not even possible to put it on a leash and show it off around town. It is God's creature and praises God in its being. It is not yours in the sense some of your pets are. This creature has strength that would overwhelm you, and it has a mind of its own. Respect this hippo, man and woman! It is God's marvelous and dynamic creation!

The hippopotamus, or Behemoth, is a remarkable creature in its own right. A beast need not have utility to have worth. Its value is in the eyes of God. God declares it to be good, as God does the whole creation (Genesis 1). It is God's property, God's masterpiece; only the Master has the right to destroy what he has created. Here we have a beast that is beautiful in the eyes of God. In our imaging God and God's love for the creation, these creatures should be beautiful in our eyes, too. If beauty is in the eye of the beholder, then we, imaging God, should see all creatures through God's eyes, beholding the beauty seen by its divine beholder.

Could Noah Afford the Ark?

Now we ask again, Can we afford it? Can we afford to keep Behemoth? Could Noah afford the ark? With a great catastrophe about to strike, should he have not just looked out for himself and his family? It certainly would have been much less time-consuming and less expensive to build a smaller boat! And it would not have required knowledge of the requirements of all those different kinds of animals. Or if only some could be saved, why not just the useful animals? Could Noah afford the ark? Apparently he thought he could. Perhaps he saw the world through the eyes of God, the Maker of all things!

Can we afford it? From Job, God elicits respect for the Creator's awesome work. From Noah, God elicits dedicated service to save the Creator's masterpieces. From us, God seeks to elicit respect for his covenant with all life that it should not be destroyed again. The clear teaching of the Scriptures is that no expense of time, material, money, or reputation should be spared in this rescue operation. God's instructions to Noah (Gen. 6:14–16) necessitated an immense cost in time and resources. They necessitated as well a great cost in personal reputation, including the mocking derision and humiliation anyone should expect who would ever do such a thing with no apparent prospect for its usefulness.

God expected Job and Noah, and expects us, to learn from and know the things he has made.[1] The request to Noah to "take every kind of food that is to be eaten" (Gen. 6:21) indicated the need for substantial knowledge of the creatures, including their food and living requirements. Noah had to know what he was doing, achieving substantial biological knowledge of all the species he was saving. Noah also understood that God's endangered species act was insufficient by itself. When the rescue operation was completed, the species were to be released to their required habitats "so they can multiply on the earth and be fruitful" (Gen. 8:17). Their saving was followed by their restoration to habitats that would assure their return from a threatened status to the status of flourishing creatures.

In our day some claim that a proposed human activity should be permitted even when it brings about extinction of a species of plant or animal. This question was put to me recently: "We have plans to develop a golf course, but have found that it is the habitat for an endangered species; what do you think?" I answered with another question: "Is this the last remaining site for this species?" The answer was yes. I replied, "Is this the last remaining site for a golf course?" Whenever we purchase any part of God's creation, we purchase the life it supports and contains. While we are not prevented from transforming or altering this part of creation, we are responsible—as Noah was, and Adam before him—to serve the garden and to name the creatures under our care. When we buy a part of the landscape, we buy a part of God's creation that we have decided to put under our own care.

Deep down, human beings know this. They have a good sense of what the integrity of creation means, but they may seek freedom *from* stewardship of what they have put under their own care. Purchase of a part of creation, however, is a decision to seek freedom *for* stewardship. While knowing this, we still find ourselves and others engaged in degrading actions. This is the human predicament.

Can we afford to keep Behemoth? Not only can we afford it but we have a motivation for doing so that is important to share with others across the globe, especially at this critical time of crisis. We must not keep our knowledge of God's expectations to ourselves. Biblical stewardship offers hope for the world. It is good news to be shared! It is particularly good news because it has the answer to the human predicament, since it recognizes and deals with the pervasiveness of human sin. In doing so, it comes to grips with our problem of doing the things we should not do and not doing the things we should do.

Biblical stewardship also offers hope to the world by addressing the problem of people wanting always to press beyond the limits God has established in creation.[2] It addresses this, among other ways, by placing a temporal limit for us and creation—a sabbath for people, for the creatures, and the land (see Exod. 20:8–11; 23:10–13; Lev. 25–26). This boundary in time assures that the land and God's creatures are aboundingly and sustainably fruitful.[3]

Biblical stewardship also offers hope to the world by recognizing that the curse rests not on the world as such but on what is sinful in it.[4] What is bad in the world is what we should be addressing in our human action. Thus, we should be pursuers of justice in all areas of life, and we should take an unyielding stance against the destroyers of the earth. We are encouraged in this stance by the knowledge that faithful and obedient people find life and are preserved from destruction (see Gen. 6:18; 7:1). All, then, must be good and faithful stewards.

Biblical stewardship offers hope for the church by engaging the people of God in the care and keeping of creation. The message of stewardship is an invitation to join with the children of God to meet creation's eager expectation. This message is reinforced by the reality of the coming judgment, described with fearsome force in the Book of Revelation. Those who destroy the earth will also be destroyed (Rev. 11:18).

Biblical stewardship offers hope for the church by recognizing up front the problem of sin. The cause of those human actions that are uncreating the world, degrading its integrity, and abusing creation is human sin. Sin is the root of crisis. Deep down, the church understands that dealing with sin is essential to dealing with the crisis.

Biblical stewardship—acknowledging Jesus Christ as Lord of Creation—brings us, with Noah, to say, "Yes, we can save our Lord's Behemoth." Moreover, it brings us to do so out of respect for its Artist, because it is he who has given us charge over the care and keeping of the things he has made.

Conclusion

The puzzle with which we began was this: The claims of Jesus Christ on the world as Creator, Integrator, and Reconciler are largely neglected by those who follow Jesus Christ. We know that the pieces of this puzzle don't fit. The first piece says, "We honor the Great Master!" The second piece says, "We despise his great masterpieces!" When we considered that the ways of life followed by people who may not call themselves religious are functionally equivalent to religious ways of life, we opened ourselves to discovery. We unmasked our "number one" religion, our "seek me first" way of life, which contends with Christianity over who is lord of creation. Thus, the two great ifs: If the impersonal market is lord of creation, then the whole creation is merely a resource and is subjected by consumers to impious use in behalf of self. If, however, Jesus Christ is Lord of Creation, then the whole creation is revelation and is served by stewards who care for and keep it in behalf of its Creator.

Here we come to the first of our three big questions, Is Jesus Christ Lord of Creation? Our answer must be, Yes! Jesus Christ is Lord of Creation. How, then, do we explain the second piece of the puzzle, the one about despising God's masterpieces? Why is it that we find so many who love

Jesus but do not give a care for creation? Reluctantly we must look to our "number one" religion for the answer. It has been parading in a mask, and even many Christians have incorporated it into their belief system.

Huston Smith, in his comparative study of religions, has informed us that religions may yield to forces they attempt to challenge. The situation for us is this: We have so yielded to society's number one religion that we can recite its central "number one" premise as though it were biblical truth. What makes this possible is the dismembering of our concept of God into Redeemer on the one hand and Creator on the other. If this dismemberment is accompanied by our adoption of the belief of our "number one" religion that the whole of creation is resources, we then in our minds can capture our Creator as the provider of those resources. In that case Behemoth has its worth defined by the "number one" economy, as a zoo animal, a potential provider of a cure for some human disease, an attractant for eco-tourism, or whatever. Behemoth is no longer seen as a creature whose primary purpose is to bring God praise as a masterpiece of the Great Master. Behemoth no longer can be one of the creatures we speak of each Sunday when we sing, "Praise God all creatures here below."

To cover for this dismemberment of our Creator, we lessen our singing of the psalms, reduce references to creation in our hymns, and move toward ever more abstract praise. Today we are more likely to praise God abstractly "for his excellent greatness" than we are to praise God concretely as the one who "sends the cheering rain," who "to beast and bird his goodness their daily food supplies."[1]

My challenge to evangelical Christians is this: Let us recognize the dismemberment of our Creator and come anew to re-member our Creator. Having re-membered our Creator, reuniting the creative and redemptive work of our Lord, we then can ask our second big question, Is creation a lost cause? And we may respond, Definitely not!

Re-membering our Creator as the one through whom all things were made, through whom all things hold together, and through whom all things are reconciled to God, we can ask our third big question, Whom are we following when we follow Jesus Christ? And we may respond, We are following the Lord of Creation, of whom it is written: "[Christ] is the image of the invisible God, the firstborn over all creation. For by him all things were created . . . all things were created by him and for him. He is before all things, and in him all things hold together. . . . For God was pleased to have all his fullness dwell in him, and through him to reconcile to himself all things" (Col. 1:15–20).

Jesus Christ is Beautiful Savior and Lord of Creation! Acknowledging that Jesus Christ is lord of all creation means that indeed we have good news for every creature (Mark 16:15). Acknowledging that Jesus Christ is Creator, the Author, Integrator, and Harmonizer of the whole creation, we indeed will meet the expectation of a creation "that is waiting . . . waiting on tippy toes . . . waiting with neck outstretched . . . for the coming of the Children of God" (Rom. 8:19).[2]

As Abraham Kuyper said, we should accept Jesus Christ whom God has given to save the world, which we have upset and broken, for it is his own "work of art," and he will "polish it again to new luster. . . . And therefore whoever would be saved with that world, as God loves it, let him accept the Son, Whom God has given to that world, in order to save the world. Let him not continue standing afar off, let him not hesitate."[3]

Environmental Realism

Response by Richard A. Baer Jr.

Just over thirty years ago I published my first article on environmental ethics in the *Christian Century,* and nineteen years ago I taught a summer course at Fuller Seminary on religion and the environment. It is gratifying to see that Christians are gradually—perhaps much too gradually—waking up to the importance of treating God's creation with greater care, and I am honored to have been asked to respond to Calvin DeWitt's challenging presentation.

Thinking about God

How can I possibly do justice to DeWitt's paper? It is comprehensive in scope but at the same time full of interesting

Richard A. Baer Jr. is professor of environmental ethics at Cornell University and an adviser to the Center for Public Justice. He has published dozens of articles and essays on the environment, education reform, and other topics in theological and scientific journals and in the *Wall Street Journal, Education Week,* and other popular media.

and significant details. His paper is both prose and poetry, and reflects the wonderful breadth and depth of his poetic imagination and intimate knowledge of nature and the Scriptures. It is also prophetic, in the best biblical sense of that term, persuasively spelling out for us God's will for human beings in relation to all other parts of creation.

Still, we must also think about how to translate these powerful ideas about God and creation into concrete environmental policy. In that connection, I am reminded of something my daughter Becky said when she was barely three years old. She had pretty well completed toilet training, but one particular day she had three "accidents," one after the other. Exasperated, her mother finally blurted out, "Becky, what's happening? Why didn't you tell me you needed to go to the bathroom?" Becky looked up at her mother with her wonderful big brown eyes, thought for a moment, and responded, "Oh, Mommy, Mommy, I was thinking about God." I am happy to report that Becky is still thinking about God. I even dare to believe that her working for a Ph.D. in religion and social ethics will not prove to be too much of a hindrance to such thoughts.

DeWitt also does a lot of thinking about God, and that is all to the good. We would all be better off if we thought more about God. In the realm of environmental policy, however, we must also concern ourselves with translating such thoughts into specific recommendations for action. Thinking is to be followed by obedience. DeWitt poignantly asks, "Can we afford to keep the Lord's Behemoth?" and answers with a resounding yes. "The clear teaching of the Scriptures," he claims, "is that no expense of time, material, money, or reputation should be spared in this [Noah's] rescue operation."

Is it really possible, however, to move so quickly from thoughts about God and Noah's ark directly to species-preservation policy? I find this problematic, about as problematic as basing food policy today on the biblical accounts of God's providing manna in the wilderness or Jesus' feeding of the five thousand. Scientists tell us that 99 percent or more of all species that have existed over time are now extinct, and almost all of these creatures disappeared long before humans appeared on the planet. If each species is as valuable to God as DeWitt implies, why was God apparently so unconcerned about their survival? I do not ask this question irreverently. Are we as Christians really supposed to preserve species at all costs? What does that even mean? How are we to balance human needs against the very substantial sacrifices entailed in protecting individual species? Are some species more valuable than others and thus deserving of greater preservation efforts? And if we simply want to preserve a large number of species, would it not make sense to abandon some of our very costly efforts here at home and concentrate on preserving tropical rain forests, which contain a far greater variety of species per square mile than temperate climate lands?

Realism about Creation's Diversity

I may not be able to present convincing arguments to support the claim, but it is my belief that to lose a dozen species like the elephant, the Bengal tiger, the mountain gorilla, and the dolphin would be a far greater tragedy than to lose ninety or even ninety-nine of the roughly one hundred species of snail darter. All species are valuable, but perhaps even God values some more than others. The incredible complexity, beauty, and ability of the so-called charismatic megafauna to fascinate and delight us as human beings must count for something in making moral judgments regarding preservation priorities. Anthropocentrism is wrong, but that does not mean we should not take our own interests into account at all.

The Bible clearly presents human beings as specially loved by God, and it seems to me not unlikely that God values some of his nonhuman creatures more than others.

DeWitt says as much in his discussion of Behemoth. If God prefers some of his creatures over others, why is it wrong for us to do so? Although the Bible presents redemption as eventually including all of creation, it is clear that the biblical narratives focus primarily on human beings, on those whom God has predestined to become his children through repentance and faith. God pronounces all of creation to be good, but it is by no means clear that our Christian commitments make it wrong for us to improve our own situation, even when this exacts a heavy cost from other creatures, perhaps even causing certain species to become extinct. DDT may have put undesirable pressures on particular species of wildlife, but it has at the same time saved tens of millions of human lives by making it possible to control malaria. I find it hard to believe that our Christian commitments are such that we should have favored altogether abandoning the use of DDT the moment we understood its deleterious effects on particular animal species.

Technology and market economies also have often resulted in damage to our natural environment, but two things need to be said by Christians in response. First, on the whole, market economies have been far more successful in moderating the negative effects of technology than have centrally planned economies. Second, life in antiquity and the Middle Ages was not really all that wonderful. Famine, plague, and sundry natural disasters regularly ravaged human communities. Indeed, in our own country, as late as 1872, Philadelphia lost close to 2,600 people to smallpox, and in Memphis, Tennessee, 8,000 people died of yellow fever in the decade of the 1870s.

When the ancient Hebrews desacralized nature by dethroning the pagan nature gods and when the Ionian Greek philosophers began to ask about the material causes of things, they removed important obstacles to the development of modern science. So long as nature was seen as partaking of divinity, it was unlikely that modern scientific methods could have developed, for they would have been seen as essentially sacrilegious. Unfortunately, a desacralized nature is also a nature that may more easily be brutalized than a nature that is essentially personal and pervaded with divinity. The Hebrews did not so brutalize nature, because they understood nature as God's good creation, and themselves as God's stewards of that creation. They knew they were to exercise dominion over nature, but always under the authority of God. As DeWitt correctly points out, they understood themselves to be servants rather than lords with respect to nature.

When modern secularists reject God's authority and replace the servant model with the image of autonomous humans committed to using reason to maximize their own self-interest, the image of humans exercising dominion over nature becomes a scary image indeed. Actually, to say that human beings have dominion over nature is not to make a moral claim at all; it is only to state what is empirically true. Whether or not we as human beings want to, we will inevitably exercise enormous power over nature. That is simply part of what it means to be human. The only interesting question is how this power will be exercised.

DeWitt describes economic markets as a gift of God but a dangerous gift. I agree. Again, realism demands that we ask, What alternatives to markets do we have? As bad as they may be in some respects, almost all the evidence we have today shows they are extremely effective in producing wealth, and it is important to remind ourselves that conservation and preservation efforts generally are features of wealthy rather than poor societies. With the collapse of centralized planning in the Soviet Union and China, the argument over whether centralized economic planning is preferable to free markets has been settled. The interesting questions today have rather to do with how we can guarantee that markets take proper account of environmental concerns and how government can assure distributive justice for all citizens.

Premodern societies did not ravage nature to the extent we have done in our century, but this was in large part because of low population levels. Because they did not have the advantages of modern medicine, water treatment, and effective sewer systems, their birth and death rates remained more or less in balance. The best available evidence suggests that they exploited their natural resources roughly up to the limits of their technological capabilities, but there just were not enough of them to damage nature in the way poor people are doing today in parts of Africa and Asia. Third World nations are more likely to come to peace with nature by becoming richer than by consuming less and remaining poor, and economic markets seem to be the best bet for creating wealth.

Sphere Sovereignty and Moral Judgment

Kuyper's emphasis on sphere sovereignty can help us at this point. Markets are not capable of doing the job of governments. Markets are extremely proficient at creating wealth, but it is government's job to prevent the formation of monopolies that distort market decisions and to make sure that markets internalize environmental externalities in the costs of production. Government should also serve its citizens by insuring that the distribution of wealth is just. Liberal economist Charles L. Schultze has persuasively argued in *The Public Use of Private Interest* (Washington: The Brookings Institution, 1977) that economic incentives sometimes work far more effectively and efficiently in protecting the environment than do regulations promulgated by government bureaucrats. Rather than endlessly multiplying regulations about how much to insulate refrigerators or what gas-mileage standards to require for automobiles, it probably would be far more efficient simply to raise energy prices to roughly European levels. When gasoline sells for three to four dollars a gallon, people will consume less of it. Triple the price of electricity, and appliances will become more efficient. Of course, justice demands that adjustments be made to protect the poor from the regressive effects of such excise taxes. Likewise, other taxes would have to be cut proportionately so we do not end up with more government and less overall economic efficiency.

DeWitt might also have applied Kuyper's thinking on sphere sovereignty to the role schools and universities could play in bringing about a healthier relation to nature. The problem with our present educational system is that it fails to distinguish adequately between the role of the state, whose task is governing, and the role of the school, whose task is educating. In a pluralistic society such as our own, the state never can educate in a just manner, for either it will sponsor curricula that are ostensibly neutral among worldview alternatives but at the same time utterly bland and largely incoherent or it will present a more coherent and forceful curriculum that will be inescapably oppressive to various religious and cognitive minorities. One school and one curriculum for all does not work well in a pluralistic society.

A properly differentiated system would permit schools to be genuine schools, not mainly state instruments for social control. It would do this by properly differentiating between educating and governing. The governing task inevitably is monopolistic: The state rules over all and ought to have a monopoly on the use of force. But schools need not be uniform and monopolistic. We can allow different groups to educate their children in different ways. Such a differentiated educational system would permit individual schools effectively to inculcate morality and foster character development in students, in part because they would be able to be thick rather than thin moral communities. That is, they could integrate moral education with celebration, worship, the use of symbols and common narratives, and so forth. Teachers could function effectively as role models, for schools would be free to hire and fire teachers in part on the basis of how well their personal lives modeled what they were trying to teach. In differentiated schools there would be no need for the endless bickering over curriculum that inevitably occurs in a winner-takes-all system. Schools in a differentiated, pluralist system would be free openly to challenge what DeWitt refers to as "our number one religion" of self-centered consumerism. They could do this not just by discussing Christianity in an "objective," descriptive manner but, if they chose, by modeling at many levels what it means to be committed to Jesus Christ and his church.

Christian schools could help children learn to appreciate and even fall in love with nature, thus pulling them out of their preoccupation with themselves. Such schools would at the same time help students understand that nature is never worthy of ultimate loyalty. God alone is Lord. Building on Reinhold Niebuhr's understanding of insecurity and anxiety, Christian schools could also help students understand why greed and exploitation of nature are deeply spiritual issues. As Niebuhr so powerfully argues in *The Nature and Destiny of Man*, we are insecure because we are finite, subject to diminishment and death. Animals are also insecure, but because we, unlike animals, are free and can transcend ourselves to some degree, we become anxious. Anxiety is not itself sin but its precondition. Anxiety is not sin because there always remains the possibility that we can trust God to provide for our present and future needs. When we fail to trust God, however, we try to establish our own security, usually by exploiting other people and our natural environment. Our futile attempts to justify ourselves and secure our own lives are always at the expense of other life. Such theological analysis makes it clear that our relation to nature, just like our relation with fellow humans, is deeply influenced by our faith, or lack of faith, in God.

Caution, Not Disagreement

Overall, DeWitt asks the right questions. My comments should be understood more as a word of caution than as a deep disagreement with him. Yes, it is wrong for Christians to view nature only in utilitarian and exploitative terms. Yes, we insult God when we insult his creation. Yes, we ought to try our best to keep the Lord's Behemoth. On the other hand, we need to be wary of those forms of environmentalism that treat nature as ultimate reality and human beings as a blight on this fragile planet. We need to be especially on guard against those environmental educators who insist on pressing a kind of neopaganism or pantheism on public school children in the name of saving the earth. Many of the values of these radical environmentalists are subversive of Christian faith and antithetical to the gospel. Animals are not as valuable as human beings. Contrary to most radical environmentalists, human beings are not just like all other animals. We are not simply part of nature; we transcend nature in that we can share in a conscious relation with the God who is transcendent as well as immanent.

In terms of environmental policy, we will at times have to form coalitions with those who hold beliefs sharply at odds with our Christian commitments. Because God values nature for its own sake and because human beings will suffer if we mistreat nature, we must radically modify how we live and think. Balancing human interests and the interests of nonhuman nature will demand hard thinking, but it is imperative that we remain faithful to all aspects of our tradition. Carl Sagan notwithstanding, the cosmos is not "all that is or ever was or ever will be." Aldo Leopold simply got it wrong when he tried to reduce ethics to the statement that "a thing is right when it tends to preserve the integrity, stability, and beauty of the biotic community. It is wrong when it tends otherwise."[1] His heart may have been in the right place, but taken at face value such a principle is too severe in its discounting of distinctly human interests.

Can we afford to keep Behemoth, that great and wonderful creature? I hope so. Indeed, I am confident we can do so if we put our minds to it. Can we afford to preserve all of roughly one hundred species of snail darter? Perhaps, but as a Christian, I would not be willing to sacrifice much in human well-being to do so. Do Christians have simple and definite answers to questions of species preservation and other complex environmental issues? Obviously not, but at least this much is clear: DeWitt has pushed us in the right direction, and we are all deeply indebted to him. We have important choices to make, and his thoughtful treatment of these matters will undoubtedly help us make those choices in a manner more consonant with our commitment to Jesus Christ as Savior and Lord.

The Complexity and Ambiguity of Environmental Stewardship

Response by Thomas Sieger Derr

There is much to like in Calvin DeWitt's paper, with its strong and unabashed Christian piety and the vigor and attractiveness with which it is argued. Given so much faithful erudition and expository charm, it might seem ungracious of me to differ (even though that is my assignment), so I will try to sneak up with my own views so quietly I won't shock anyone.

The backbone of DeWitt's analysis is straight and clear: The earth is in ecological crisis, or "geo-crisis"; much of that has been caused by human greed; the scientists are sounding the alarm; and religion, if it wakes up, can and ought to tip the balance.

I have to confess that I see the situation as somewhat more complex and ambiguous. When I survey the vast liter-

Thomas Sieger Derr is professor of religion at Smith College. He has authored numerous books and articles on the environment, including *Creation at Risk? Religion, Science, and Environmentalism* (Grand Rapids: Eerdmans, 1995) and *Environmental Ethics and Christian Humanism* (Nashville: Abingdon Press, 1996).

ature, I find discord among the scientists. When I look at the human impact on the earth, I find a wide variety of practices and motives, so I cannot say rapaciousness reigns everywhere, though it undoubtedly does in some instances. When I look at the natural world, I find that it is not without evil; indeed, I would say that creation is, theologically speaking, "fallen," that theodicy is still the major theological problem that in one way or another bedevils every Christian and is sooner or later a theme of every pastor's preaching. Finally, when I read religious writings on environmental problems, I find to my distress that they can all too easily come up with the wrong solutions, that they frequently do not provide the answer after all.

Reading the Scientists

I do not want to commit the error of asking someone to write a paper other than the one he set out to write, so I do not fault DeWitt for not discussing these matters. But I would like to discuss them. To take them in order, I would first of all admit to real puzzlement when I read the scientists on these matters. Any lay reader who goes far enough into the literature will find the absence of consensus striking. Take the famous global-warming controversy, for example, which comes close to being the crisis of choice these days.

When I was younger, I remember clearly that global *cooling* was predicted, that the snows were mounting in the Arctic and we were likely to be in for another ice age, or at least a "little ice age" like that experienced from the sixteenth to the eighteenth centuries. Since I am not as old as Methuselah, this alarm was really rather recent—about thirty years ago. Climate changes move over the long term, and recent data thus tend to be unreliable. I will not argue for either side of this debate; I do not know enough. But I certainly get discouraged, as a layperson, when I see my scientific betters, sporting their Ph.D.'s and their prestigious appointments, squabbling over the meaning of their measurements. There is, to put it simply, a difference of views among climatologists as to whether there is any meaningful long-term global warming going on and, if there is, whether human activity has anything to do with it. Those who want a quick glimpse into the passion of disagreement can look at the August (1996) issue of *Physics Today,* where the quarrel is reported with admirable objectivity. It seems that a report by the Intergovernmental Panel on Climate Change was altered in the editorial process to imply more clearly that there is an anthropogenic factor in global warming. Authors of the report claim normal editorial rewriting; critics descry intent to deceive, and the fur flies.

Climate change is just one issue. We could document severe disagreement on the ozone hole, pollution, the loss of fertile lands and forests, loss of species, exhaustion of resources, acid rain, nuclear energy and waste, and even population growth. And as we have learned through a long and sometimes painful history, making judgments on the conclusions of natural science is not an area in which a specific Christian conclusion is readily available. How, then, are we, as citizens, members of the public, and Christians, to proceed in the face of these disagreements? I suggest some brief guidelines for the Christian conscience.

To start with the obvious, we should demand that partisans adhere to what we old-fashioned moralists think of as good character: that decisions should come from cool-headed reason and honesty, laced with a willingness to weigh charitably the other side and not indulge in deprecatory remarks or name-calling. In a book published just recently, *The Betrayal of Science and Reason: How Environmental Anti-Science Threatens Our Future* (Washington: Island Press, 1996), Paul Ehrlich, frequently a target of those who think his environmental apocalypticism is wrong, fires back at this criticism as "antiscience" and "brownlash," stating that most of it is bad science, and comparing his critics to "scientific pretenders with perpetual motion machines, flat earth theories, and creationist beliefs." I should think we could do better than this if we are to maintain the kind of discourse that might lead to sound public policy.

My second guideline is to be cautious about choosing sides based on our like or dislike of the messenger, though inevitably some assessment of personal credibility will enter in. We have a tendency to associate environmentalism with one side of the political spectrum and the opposition with the other. Or we associate businesspeople with "antienvironmentalism" and judge the science at issue on the basis of our own attitudes toward businesspeople. None of this is really relevant, of course, and we would do much better to concentrate on the science. The facts do not depend on whether they fit the agenda of either liberals or conservatives, in any sense of those overused words.

My third guideline—and I think on this one we can claim support from Christian realism about human nature—is to allow for the effect of self-interest on the rival claims to truth, again sorting out the science from the nonscientific factors. Of course producers and users of harmful substances want to protect their economic interests and may hire researchers to bolster their claims to environmental innocence. This warrants an especially close look at any findings of such provenance. Agencies whose continued funding depends on a sense of environmental crisis may campaign for funds by press release of speculative and frightening data, and we should ask that their results be peer-reviewed rather than manufactured for political purposes. Private environmental organizations themselves depend for their very existence on giving the impression that they are saving the planet and its creatures, and we should not ignore their self-interest in assessing their reports. I do not mean that these reports, claims, and findings are fatally compromised; on the contrary, they are to be taken seriously. But we should make every effort to detach the facts from any slanted presentation.

Fourth, I suggest simply that we not treat environmental disputes like straight-ticket party voting. It is not the case that one "side" is right and the other wrong but that there are multiple opinions on many issues. Maybe the earth-warming scare is overblown but the ozone-hole alarm is right, for example. We will have to pick our way through, issue by issue.

Finally, I would have us look into ourselves to discover, if we can, the effects of our own makeup on our judgments. If we are worriers, naturally given to fears, we may too easily fall prey to worst-case scenarios. If we choose according to what we want to be the case, on the other hand, we may become complacent about dire environmental predictions and be unwilling to make allowance for their possibly coming true. Knowing ourselves is a crucial piece of the judgment process.

To sum up this first major point of my response, I would have to favor not a program for action based on scientific certainty that the crisis is upon us but a program designed for uncertainty. This is not nearly as comforting and involves us in the ambiguous business of risk assessment.

Public Policy Concerns

The second section of my response is directed to the public-policy questions. Awareness of environmental problems does not lead automatically to one political approach. It is true that many of the most apocalyptic alarmists take a radical approach to public policy, which one of their critics, Ronald Bailey, has labeled "millennialist." We must adopt wholly new ways of living, they say, and had better start right now with government leading the way. Regulation, planning, and central authority are needed, nationally and even internationally, to restrain the negative ecological impact of private, selfish decisions and actions. In short, in the perennial tug between freedom and order, this group of environmentalists has opted for order.

This is the sort of environmental politics that tends to be suspicious of capitalism and the market economy, but there are other kinds more favorable to the economic system we know in this country. One such approach is what we are doing now and represents the state of current public policy: Use tax incentives and user charges to bring about improvements piecemeal. Polluters will clean up their act if effluent charges make pollution costly, and they will add the cleanup costs to the price of their product, as is proper in a market economy. Tax breaks are used to offset the lost opportunity cost of not developing wildland, combining the self-interest of owners with the desired environmental goal. Direct regulation has a place here, but mainly as a backup, to accomplish what more positive incentives cannot do as easily.

There is another approach that takes the usefulness of self-interest to its logical conclusion and does away with regulation altogether. This is the so-called free-market environmentalism, whose theory is that if we privatize resources and allow their exchange in the market, their owners will preserve them, practicing the sustainability that is to the ecological benefit of all. After all, one does not destroy one's own productive capital. Proponents do not like the gradualist approach of tax and fee incentives, because there the government is still determining the environmental goals. Better to leave these goals to the free determination of free individuals, lest in matters ecological we fall back into the errors of central state authority from which many countries are only now emerging.

Free-market environmentalists may themselves be criticized for their too easy assumption that resource owners satisfying private wants will automatically serve the common good. Owners may instead run a resource down, take their profits, and invest them elsewhere in the economy. If value were defined only by market price, wetlands would be drained for land development even if they were ecologically essential. Moreover, such a pure free-market approach does not give any voice to those who have no capital but must live in the environment produced by owners' decisions. These criticisms, incidentally, are also made by DeWitt and his colleagues in the book *Earthkeeping,* and again in this paper where he argues rather strongly that unrestrained market values are our "number one religion."

Faced with these competing social visions, the Christian conscience is not without resources, though a definitive answer is not supplied by our faith, either. We know that we must serve both the needs of individuals and of the community and that some kind of balance between them must be found. Stressing the former, we will opt for policies favoring private decisions, consensual exchanges in a free market, and freedom from government edict. We will remember the Christian principle of subsidiarity, that decisions should be made not centrally but at the lowest level possible, so people may have some real control over their lives. To love one's neighbor means to ask what the neighbor's need is, not to define it for him.

On the other hand, Christians have always stressed service for the common good, which is explicitly not the sum of private wants but the good of the community as a whole. Thinking in this mode, we will be wary of private selfishness, remembering that rights to ownership and use have always been qualified in the Christian tradition by the requirements of the general welfare.

Love in the Service of Stewardship

I cannot pretend to resolve the tensions here, for they are permanent. I know that moral suasion has its limits and that an adroit use of self-interest can be harnessed to the common good. At the same time, however, we cannot give up on appeals to selflessness. Others-regarding love, *agape,* is, after all, the Christian behavioral norm, even if we sinful humans compromise it all the time. Thus, I am fully in accord with DeWitt's appeal to selfless love in the service of stewardship of the earth. I simply want to bring out some of the tensions and ambiguities involved in translating this ethical imperative into public policy.

I have no argument whatsoever with DeWitt that Christians must affirm the creation as a matter of faith in the goodness of God. But the problem of evil remains; indeed, it is the goodness of God that gives the problem its poignancy, as we all know. Still, the creation is not the Creator, and biblical faith is not pantheism or monism. The possibility for evil, or fallenness, is thus introduced. This is not an evil that we can ascribe to human sin in any direct sense, for what occurs in nature that harms us—for example, plague, drought, volcanic eruption, or violent storm—is not normally of our doing. The world of nature sustains our life, but it also takes that life away. May we still call it unreservedly good?

Many people, I among them, agree with Saint Paul that despite the goodness of its Maker, the entire creation sighs and throbs with pain (Rom. 8:22). Some would go on to say, as Paul does, that it awaits its redemption. The theology of the Orthodox Church does that, applying its doctrine of "theosis," or divinization, not only to human destiny but to the destiny of the earth as well. Their argument is buttressed by the claim that the incarnation confers divine dignity not only on humanity but on materiality itself, all destined for salvation in the death of Christ. This notion of cosmic redemption, as it is sometimes called, has been seized upon by many theologically minded environmentalists outside the Orthodox Church to validate the ultimate goodness of the natural world despite its often harmful impact on human beings, and DeWitt associates himself firmly with this larger school of thought.

I would be much more cautious. I think of the natural world as curiously veiled in its goodness and quite provisional. I cannot bring myself to believe that the earth will be spared its final destruction when our sun expires four billion years hence. Nor can I allow myself even to speculate on what will have become of the human race by then. I do believe that salvation ought to be construed primarily as a present reality, in and for the world we live in. But the ultimate destiny of our world, our sun, our galaxy, our cosmos, is beyond our ken. I will put my trust in the goodness of God to bring all to resolution, and I will consign the details to silence.

Unlike me, DeWitt seems to partake of the optimism about nature that was typical of the Enlightenment, although he has expressed it in an altogether different vocabulary, that of evangelical Christianity. As I listen to him describe the world as admirably arranged on a "human scale," crediting the Maker with allowing that which is not of human scale—for example, the dinosaurs—to vanish, I am reminded of Leibnitz, who thought that this was "the best of all possible worlds" and whose solution to the problem of theodicy (a term he is credited with coining, though the problem is ancient and eternal) was that natural "evil" was really like the dark shadings in a picture, there to bring out the good.

I wonder instead—and this is just speculative musing—whether nature is prearranged only in the more general sense that it displays intelligent design but in the details has a certain freedom. If it had a will, I would say it had freedom to sin. But since I believe it is without will and is thus amoral (not immoral), I will say tentatively—without, of course, any way of really knowing—that nature's freedom inheres in its randomness, a freedom that many scientists think they find when they study the natural world.

I ask myself whether there is a further problem here, not only an optimism that I cannot fully share but also a touch of determinism that what is ought to be. "Natural kinds are good kinds until proven otherwise," says Holmes Rolston, philosopher of nature and Presbyterian minister. "This biological world that *is,* also *ought* to be; we must argue from the natural to the moral."[1] I do not want to put words in De-Witt's mouth, and this is not a quote from him, but I think it represents his general view, and it worries me. Fearing fatalism, I have a more active, interventionist view of the human responsibility toward creation.

It may be a matter of some curiosity that I should disagree with DeWitt in my appraisal of nature, that I should be more pessimistic. After all, I too have been raised in the Reformed tradition, and I've been ordained to the ministry in that tradition. And although I am not a scientist, I was certainly raised in a scientific culture: Both my father and grandfather taught physics at MIT, and my other grandfather was a doctor. So how do I come to see the world rather differently? Perhaps my experience of nature has been different, or perhaps something in my temperament has reacted differently to nature's pressure, which I have learned to think of as amoral and, in its effect on me, as much hostile as nourishing. It is mindless, amoral nature whose leukemia killed my father and my sisters when I was young. It is randomly heedless nature whose chance mutations gave one of my sons clubfeet, and successfully defiant human medicine that saved him from the life of misery that nature "intended," if intention be allowed. So biography has its effect on theology, as it always has, and perhaps we should let it go at that.

Religious Responses

There is another way to look at the ambiguity of nature—the biocentric solution—and this will bring me to my last point: namely, that religious responses to environmental problems are not always helpful. Biocentrism argues that the whole life process is good and to be valued in its entirety. Species and ecosystems take priority over individuals. Indeed, the life process wills the death of individuals so that the whole may survive. It is even natural that species themselves perish and be replaced by others, and that is not wrong so long as we humans do not interfere with the process by causing the extinctions ourselves.

There is an implicit—sometimes explicit—theological assumption here: that value resides in the whole, the entirety, and that this natural whole is good no matter the fate of individuals and species within it. It does not matter that the human race will one day follow other species into extinction, either, as long as the life process goes on. Religiously minded people who have bought into this worldview are able to praise its Maker, if they are monotheists, or praise the cosmos itself, if they are pantheists or monists, without being ethically troubled by the prospect of human demise. Among them are people who call themselves Christian biocentrists, and as you can see, they have solved the problem of natural evil rather neatly: There is no such thing. It is a false perception on our part that the harm nature does us is evil. That would be an anthropocentric way of looking at the world, and we should forswear such a viewpoint as environmentally disastrous. It is such human-centeredness that has brought us to the point where our own species is a threat to the life process itself, and we should forthwith announce our conversion to bio-centeredness, or biocentrism.

I would rather frankly embrace the term anthropocentrism as an implication of our creation "in the image of God," and I am willing to defend it even against those anti-Christians who think the *imago Dei* is the root of our ecological troubles. As a modern "Christian humanist," I would bend our learning and skill to the service of human survival, and this is the source of my theological environmentalism.

I am not sure whether such a stance puts me at odds with DeWitt, for I am in accord with most of his practical conclusions. I fully agree that our environmental situation calls for the exercise of faithful discipleship in the stewardship of creation, which is the exercise of our proper human dominion. I also agree fully that dominion is not domination, despite the way anti-Christian environmental philosophers and historians have read Genesis 1:28. This is a point I have made repeatedly in my own writings. Our obligation is responsible care for the true and ultimate Owner, and this absolutely rules out the idea that we may do as we please with creation. I also agree that this special responsibility is a consequence of our creation "in the image of God." The result of our power is responsibility.

Within this general agreement we may have some differences. Does our proper dominion mean service of nature itself, not only of God or of humankind? And may one appeal to the example of Jesus to say so? DeWitt suggests that in his paper, and the book *Earthkeeping*, on which he collaborated, says so plainly. I wonder if it is not a stretch to say that "Jesus Christ brings us to see this dominion as service" to nature. Philippians 2:5–8 is not about service of nature. We should be cautious at this point lest we ascribe obligations or even "rights" to nature that could require priority over human welfare.

I would stress again the point I made above, that we are made to cultivate and manage the earth, not accept passively whatever nature brings. The "integrity of creation" is a slippery and deceptive standard, given the dramatic variability of earth over geological time, and as a slogan I do not think it gives us much of a standard. A quotation in DeWitt's paper says that we ought to hand on the earth to our descendants as we received it "or better cultivated." That would be my point: managerial stewardship. Let us beware, then, of those who find static nature the locus of supreme value, even though many of them are religious people.

I would also be careful about the ecological implications of the story of Noah and the flood. Noah kept only breeding pairs. The rest, presumably innocent, were left to drown. In the story, the people left to drown were emphatically not innocent. We must beware of the parallel here, for it means that species should be preserved at the expense of individuals. It is important, therefore, *not* to draw parallels between our treatment of the natural world and our treatment of humanity. The antihuman dangers are not far away if we do not keep the distinction clear.

Let me conclude with a word of appreciation. I was asked to seek out contrasts in order to stimulate dialogue, an assignment that has perhaps given these remarks more of an oppositional character than I meant. I do not want to make too much of what may after all be only differences of emphasis between us. In its larger themes DeWitt's paper is certainly on target and important to heed. I only want to make the case for a human-centered approach to environmental problems and to argue that that is what Christian stewardship requires of us.

Christians, Politics, and the Environment

Response by Vernon J. Ehlers

Evangelical Christians owe Calvin DeWitt a great deal, not only for this paper but even more so for his energetic activities on behalf of the environment and the Christian's responsibility to God's world. He is not only an academic leader but also an activist and a teacher. His words and works have been an inspiration for many, particularly students, throughout the years.

Dr. DeWitt has outlined a persuasive case that Christians have a deep responsibility to care for this planet and all its flora and fauna. Perhaps the most important message we can glean from his work is that dominion, as outlined in Genesis, can be understood only in terms of service.

Vernon J. Ehlers (R-Mich.) was first elected to Congress in 1993. The first research physicist in Congress, Ehlers quickly became the vice chairman of the House Science Committee. Prior to entering Congress, Ehlers served several terms in both the Michigan House and Senate. He taught for years at Calvin College before entering politics full-time and he is coauthor of *Earthkeeping in the Nineties: Stewardship of Creation* (Grand Rapids: Eerdmans, 1990).

Dominion as service must be the guidepost for all people as they interact with the ecosystem and seek to fulfill the commands given in Genesis 1 and 2.

There are issues I wish DeWitt had mentioned. It is important, for example, to look at the implications of justice as it applies to environmental problems; the impact of population growth on environmental problems; and the structural aspects of our economic and political systems—all of which make it difficult to achieve a change in attitude about the protection of our environment.

Religious and Historical Aspects of Environmental Issues

It is certainly true that people's religious beliefs play a major role in their attitude toward the ecosystem as well as in their treatment of the biosphere. This fact first attracted broad public attention when Lynn White wrote his article "The Historic Roots of our Ecological Crisis" (*Science* 155 [1967]) more than three decades ago. Although I believe much of White's analysis is flawed and in some cases unfair, one major point is clear: The public's attitude toward the environment is strongly shaped by religious beliefs. Another factor that emerges is that an individual's beliefs about the environment are frequently at odds with the theological interpretation one would expect based on the individual's overall religious beliefs.

As evangelicals, we can be grateful that a proper Christian (and for that matter, Jewish) interpretation of Scripture leads to the strong conclusion that we have a major responsibility not only to use the resources of the earth but also to maintain and preserve them as we till and keep the Lord's garden. It is indeed a puzzle why this was not recognized by the Christian community until after environmental problems became serious enough that they literally screamed for a re-evaluation of our attitudes and beliefs. This was first pointed out and examined in a comprehensive way in a project conducted by the first group of Fellows at the Calvin College Center for Christian Scholarship. DeWitt and I, along with several others, were privileged to participate in this work. Since that time there have been numerous other articles and books expanding and enlarging upon our themes, superseding the work we did.

In his paper DeWitt rightly points out that the "number one" religion, which is certainly antithetical to the Christian faith, has been at fault for some wrongful attitudes toward the environment. Yet I believe he is wrong to imply that this is the major religious belief behind our problems. The "number one" religion to which he refers is a passing phenomenon—a fad of the eighties—and already fading into oblivion. His point that alternative religious views create some of our difficulties in environmental work today is certainly valid. The problem, however, is much broader than the "watch out for number one" article of faith, which is only one manifestation of the problem.

As an example, one can readily argue, as Will Herberg does in *Protestant, Catholic, Jew* (Chicago: Univ. of Chicago Press, 1983), that the main religion in America is civil religion. As Herberg describes it, civil religion is the compilation of patriotic, religious, and pseudoreligious beliefs common to virtually all Americans. Although American civil religion was greatly weakened by the Vietnam War, it is certainly still a major force in terms of the common cultural practices of the American people. The attitude of "my country—right or wrong" is still evident in the intrinsic belief that America is not only great but also right, that we are specially blessed by the Lord and therefore are carrying out God's will in all that we do. Because we as Americans have been blessed with so much, we also have come to believe that we have the right to use so much; because we are the best, that we have the right to do as we wish; because we are so smart, that we can treat the earth as we wish because certainly we will be able to find solutions to any environmental problems; and because the Lord has always blessed us, that God will certainly continue to bless us with answers to our resource shortages and our degradation of the environment.

Another problem is scientism. Many people have a great faith in the scientific method, not only as a means of ascertaining truth but also as the means to finding the answer to any problem. Over the years, I have had countless individuals tell me that we need not worry about loss of wetlands, rapid consumption of fuel resources, disappearance of species, and other problems, because science will certainly find a solution to these problems. Long before we run out of oil, they believe, scientists will have found alternative energy sources that will provide us even cheaper, more abundant energy.

Another religious belief that causes problems is something I shall refer to as "environmental pantheism." This view sees god in everything, and as a result it sees God in nothing. The extreme version (the Gaia hypothesis) even regards the earth itself as a god. According to this belief, one should not harm any creature or any plant, because the earth, as it and its inhabitants have evolved, is perfect—that is, it has achieved the natural balance among all competing forces, and humans are upsetting the balance and must be restrained. The answer to all our problems is to step back and let nature take its course, making certain that we have zero impact upon the environment. Unfortunately, this view ignores the biblical view that humans have responsible dominion. The best we can say is that perhaps this is a reaction against the years in which Christians acted as if they had total dominion and no need to serve, till, or keep the earth. The environmental pantheist believes that we must only serve and that we have no further responsibility in terms of dominion or using the earth's resources to meet the needs of the people on this earth. Some have publicly stated that we should never cut another tree in the forest nor attempt to harvest fallen trees. Naturally, such a course would have serious consequences for the cost of providing housing and especially for the poorer members of society.

There are other variations on religious beliefs, or perversions thereof, that create problems. Clearly, human selfishness—taking things for ourselves and neglecting the needs of others—can lead to serious mistreatment of this planet and its resources. Materialism, which along with the "number one" religion is basically a form of idolatry, contributes immensely to the way we overuse resources of all sorts in the developed world. Once again, these sins are not a direct result of any particular religious beliefs but are certainly a result of our sinful human nature.

Not all attitudinal problems, however, arise from religious beliefs. Some simply arise from the accumulation of myriad attitudes and ideas over the ages; these too shape the historical roots of our ecological crises. As an example, most individuals tend to have very short time horizons. I find it amusing that I, as a congressman, am often accused of not having a planning horizon beyond the next election. First of all, that is not true, but secondly, I find that many of my constituents have far shorter planning horizons that do not extend beyond their next paycheck. In less-developed countries, the planning horizon often does not extend beyond the next meal. Very few individuals, particularly in the earlier years of their life, think beyond the immediate or consider their responsibility to future generations. Yet I believe, as a Christian, that we have as great a responsibility to future generations as we do to the current generations, and we have a clear mandate to preserve and protect this planet, maintain it as a sustainable unit, and ensure that our offspring and their children have the same resource-use opportunities that we inherited from our ancestors.

A related problem is the myopic attitude of many people regarding our resources. I recall a cartoon that appeared in *Christianity Today* some years ago, in which Adam is simply tossing to the ground the core of the original, troublesome apple that he and Eve have eaten. In response to an apparent chiding from Eve, he responds, "Don't worry. This place is so vast that we couldn't possibly ever pollute it!" This attitude has been prevalent among people ever since creation. This is particularly true in America, which has always possessed vast resources of every type. Americans generally have simply assumed that those resources will always be available for their use. We are only a century away from a frontier society, and many of our frontier attitudes still prevail.

A good solution to both the short time horizons and the myopic view of resources is to cultivate the view of earth as a spaceship. As we consider the Apollo voyages to the moon and the flights of the space shuttles, we become acutely aware that the astronauts are provided with just enough food, water, and oxygen to complete their specific journeys. Those marvelous photographs of earth from space are a jolting reminder to us that earth itself is a tiny sphere in the deep vastness of space, and while the Lord has provided us with enough water, land, and oxygen for our needs, those resources are limited by the size of the planet and by the manner in which we use the resources. Furthermore, we must come to grips with the jarring fact that although the Apollo craft and the space shuttle can always return to Mother Earth to replenish their supplies, Mother Earth has no other spaceship to which it can return to replenish its own resources. What we have is what we got at creation, and it is our task not only to use it but also to "till and keep it." Every child on this planet should be taught that, and every adult should have a worldview that includes Spaceship Earth as a major component.

Other Issues

There are other major issues that have a bearing on environmental problems and on which we should touch briefly. First of all, principles of justice, which form the underlying basis of God's commands to all peoples, come into play on environmental issues. DeWitt refers to the fact that we must preserve wetlands and argues that those individuals who own such resources must factor the preservation of wetlands into their responsibility for the property they own. Issues of justice, however, have also given birth to the property rights movement in the United States. This movement argues that any property owner who is deprived of full use of his or her property because of government regulations deserves just compensation from the government entity formulating the regulations. Unfortunately, the economic demands of such a requirement would bankrupt most governments, so the net effect is not compensation for property owners but rather the end of regulation of wetlands, sand dunes, and other environmental treasures.

Government usurpation of property rights is not new. We have zoning regulations, weed-cutting laws, side-lot requirements, and many other restrictions upon uses of property. However, environmental restrictions tend to result in greater financial loss for large-property owners than any of the previous regulations and thus have given rise to strong political opposition. What is the just answer in this situation? Clearly, compensation of the owner is just but, in most cases, financially impossible. At the same time, we must recognize that arbitrarily depriving landowners of the ability to use all of their property is a serious problem, particularly when they bought that property with the intent of using all of it and therefore face severe financial problems if they cannot. This issue deserves a thorough examination from the perspective of Christian justice, as well as from the legal standpoint. In the meantime this will be a major focus of environmental disputes in the legislative arena for years to come.

Another environmental problem not mentioned by DeWitt is the political, economic, and governmental structures that have been put in place as a result of many past religious, historical, and political beliefs. Even though we may correct our religious beliefs and correct individuals' attitudes, the old structures are still in place. Until that difficulty is addressed, we will not have much success in dealing with environmental problems. It is not enough just to change attitudes; we must also change laws. It is also not enough simply to change laws; we must also change structures created by old laws and practices. Fortunately, in a democratic society laws and attitudes generally change together, but inevitably there is a time lapse between the two, and this retards progress. We must identify and recognize the structural changes necessary to achieve our goals as we continue to educate the public about the implications of their attitudes and religious beliefs.

Perhaps the most important environmental issue we must confront is the increasing population growth. Returning to the Spaceship Earth metaphor, we all recognize that an Apollo space capsule or a shuttle vehicle has a certain designated carrying capacity. We normally fail to realize, however, that Planet Earth also has a carrying capacity, although it is certainly much less accurately defined than the capacity of a space vehicle. Furthermore, the carrying capacity of Planet Earth is highly dependent upon the standard of living we wish to attain. The higher the standard of living, the lower the population carrying capacity of the planet.

At this time the people of Planet Earth are proliferating with great abandon, particularly in less-developed countries, and at the same time the standard of living is increasing. Some even believe that the increased standard of living results from the increase in population, but I have never seen any clear evidence of that correlation. Rather, I believe the increased standard of living causes the increased population because better health care practices, better agricultural production, and purification of water reduce medical problems to the point that the population increases enormously due to lower death rates. The people of most developed countries have curtailed their reproductive rates in view of increased longevity, but that normally occurs years after lower death rates are achieved. In the developing world, which is already very densely populated, there is simply not sufficient time for the natural course of events to slow reproductive growth.

Once again, this is a difficult issue for many Christians. Some are reluctant to use any contraceptive means themselves, and many more are reluctant to suggest that others should limit the number of children they bear. Yet if we do not face this common responsibility, we will be depriving future generations of the opportunities we have enjoyed. One may wish to argue about the exact carrying capacity of the earth, but there is no question that there is an upper limit of some sort. It behooves us to recognize this and to seek to change attitudes so we can ensure that all peoples are properly fed, clothed, and cared for.

Unfortunately, abortion has come to be regarded by some nations as the answer to this problem. I believe Christians have to speak out firmly against this abomination. At the same time, I believe that this places upon us the responsibility to encourage other means of controlling reproductive rates.

Political Action

I applaud DeWitt for his efforts to influence legislation through the press conference he spearheaded last year on the Endangered Species Act. It did indeed receive a good deal of press throughout the country, although much of the attention focused not so much on the group's position on the Endangered Species Act as on the fact that conservative Christians were for once speaking out clearly in a unified fashion on environmental issues. Frankly, that oddity was more newsworthy in the eyes of some of the media than the content of what was said.

It is important for all of us to engage in greater political activity to influence the course of this nation in all spheres of life, not just environmental issues. However, it is also important to recognize that press conferences, although an important component, are only a minor part of the picture. There are literally dozens of press conferences held each day, alerting people to some major problem under consideration by Congress, and encouraging people to take action to correct that problem. I am afraid that Congress little notes nor long remembers most of these statements. To have an impact on the political process, a much more sustained and thorough approach is needed. I believe any good political action plan needs the following components.

1. Be Well Educated

It is important that all participants in the political process be well educated on the issues. State and federal legislators are given the responsibility to evaluate all sides of an issue and make the best judgment possible. To try to educate oneself on the many aspects of the multitude of issues confronting any governmental body is excruciatingly hard work. The bureaucracy has the blessing of limiting its attention to only specific aspects of certain issues, but the public expects elected officials to be able to knowledgeably address and decide upon a host of issues.

I find that most individuals coming into my office to lobby me, particularly citizen activists, are familiar with only one issue and, even worse, are familiar only with their point of view on that issue. It does little good to try to influence a legislator unless visitors have taken the time and energy to acquaint themselves with the entire issue, particularly those points of view with which they disagree. Words will have much more weight if the opposing arguments are recognized and convincingly countered. If citizens do not educate themselves to that extent, a visit to their elected official is little more than a simple expression of opinion, rather than a discussion of the problem, and is likely to have little impact.

2. Understand the Complexity of the Issue and the Legislator's Task

As I have indicated above, the task of the legislator is difficult. Those who advocate simple solutions generally do not understand the complexity of the problem and, even worse, do not understand the complexity of the legislative process or the role of elected officials. The citizen who does not make the effort to learn the simple elements of the political process but simply demands "action now" has little impact. Citizens who make the effort to learn the legislative process will understand the peculiar difficulties the representative faces on that issue. If they can help the elected official overcome those difficulties, they are more likely to be heard.

3. Work with Others

Contrary to popular belief, elected officials are indeed responsive to the people who elect them; if not, they will soon no longer be elected officials. Therefore, the more people sharing a given point of view, the greater their impact. One citizen coming in to complain about an issue is not going to be heard as clearly as a larger group, particularly if the larger group consists of persons representing many other individuals.

People say that politics makes strange bedfellows. Most people assume this applies to politicians, but in fact it applies more frequently to advocates of particular positions. If evangelical Christians wish to have an impact on the political process, they may indeed have to ally themselves with "strange bedfellows" as they seek to accomplish their specific goals. In an attempt to alleviate world hunger, for example, evangelical Christians may find themselves allied with Christians of all shades of belief, including the most liberal of the churches. They may also find themselves allied with humanists, others of non-Christian faith (such as some of the many politically active Jewish organizations), or even agnostics or atheists who happen to share the same political view on an issue. Evangelicals must abandon the idea that somehow they will be sullied by such cooperation. Effective political groups are able to concentrate on the immediate goal in front of them, seek allies for that particular view from every source possible, and then, after successfully achieving their objective, simply disband the coalition and move on to form new ones.

4. Be Politically Active

I believe it is incumbent upon all citizens to be politically involved. It is a travesty that less than half of our citizenry votes in elections. What is even worse is that only 5 percent are politically active and working on campaigns, and only 1 percent contribute to campaigns. Many people complain about campaign finance issues, but only 1 percent of the citizenry bothers to write a check to the party or candidate of their choice. In that vacuum, those who contribute are going to have an inordinate impact, for good or for ill, simply because their contribution is likely to assist that particular party or candidate.

I believe it is incumbent upon all citizens not only to vote but also to work for and contribute toward the party or candidate of their choice. This not only helps ensure that good candidates get elected but also helps establish a personal relationship with the candidates. Even more importantly, the simple activity of working in political campaigns will inform the voter in a way that no other activity can about the difficulties of the political process and the problems candidates and elected officials face. Political involvement gives a citizen a certain credibility in addressing issues, and that credibility is important to be effective in the political process.

5. Be Effective

Choose your actions carefully. Greenpeace undoubtedly gets more headlines than almost any other environmental group because of the outrageous actions its members frequently take. At the same time, I would have to rate the effectiveness of this group as among the lowest of the various environmental groups. Publicity does not equate with results. It is important to be public, but it is also important to be public in a thoughtful way, as DeWitt and the other evangelicals were in the press conference he mentions. It is important to be thoughtful, accurate, and considerate in one's public actions. Staging demonstrations in front of the factory or office of a polluter can on occasion be an effective means of attracting public attention, but if that is not followed with positive, concrete steps toward solutions, little is accomplished.

6. Be Sympathetic

I alluded earlier to the difficult task of the elected official. It is important that citizen activists be sympathetic to the problems elected officials face. It is not an easy life living in a fishbowl, having every aspect of one's personal behavior and family life subject to public scrutiny, working eighty or more hours per week, and spending many hours away from home and family. Publicly ridiculing elected officials is likely to hurt rather than help one's cause. Publicly displaying an understanding of the difficult position officials occupy and offering one's assistance is much more likely to produce positive results.

Democracy is not a spectator sport; it is a participatory activity. All of us must work together to achieve results. Those who sit on the sidelines and berate the referees and the players accomplish little. Those who become actively involved, who understand the issues, the participants, and the task of the elected official, are likely to win because they will be part of a cohesive team that can work together. It is my hope that all evangelical Christians will take their civic responsibility seriously by understanding the issues, including environmental issues, and then by becoming active team players in our efforts to achieve a better environment, a more stable political system, and above all a nation and a planet that are obedient to God's will.

Notes

Introduction

1. See Abraham Kuyper, "So God Loved the World!" in *Keep Thy Solemn Feasts,* trans. John Hendrik de Vries (Grand Rapids: Eerdmans, 1928), 70–71.

2. Abraham Kuyper, *Lectures on Calvinism: Six Lectures Delivered at Princeton University under the Auspices of the L. P. Stone Foundation* (Grand Rapids: Eerdmans, 1953), 30.

3. In the Stone Lectures, Kuyper explains this by saying that because of the two means by which we know God—the Scriptures and nature—there is no possibility that someone "who occupied himself with nature was wasting his capacities in pursuit of vain and idle things. It was perceived on the contrary, that for God's sake, our attention may not be withdrawn from the life of nature and creation" (Ibid., 120–21).

4. And yet, important as theology is, Kuyper in his Stone Lectures says that one should not limit oneself "to theology and contemplation, leaving the other sciences as of a lower character, in the hands of unbelievers; but on the contrary, looking upon it as his task to know God in *all* his works, he is conscious of having been called to fathom with all the energy of his intellect, things *terrestrial* as well as things *celestial*" (Ibid., 125).

Chapter 1: A Perplexing Puzzle in the Context of Geo-Crisis

1. The term *crisis* should be used advisedly. I agree with Ron Elsdon that in most situations "it is wrong to refer to an environmental *crisis,* since this word implies the existence in time of a sudden and decisive change, either for better or worse" (Elsdon, *Bent World: A Christian Response to the Environmental Crisis* [Downers Grove, Ill.: InterVarsity Press, 1981], 9).

2. As an indication that what we are experiencing, while being a crisis, is a continuing event on the human timescale, the categories used in the section titled

"Seven Degradations of Creation" are the same ones I have used for several years. The environmental problems of crisis remain with us, even as some things get better and other things get worse. See, for example, my "Seven Degradations of Creation," *Perspectives* (February 1989): 4–8; "Assaulting the Gallery of God: Humanity's Seven Degradations of the Earth," *Sojourners* 19, no. 2 (1990): 19–21; "Seven Degradations of Creation: Challenging the Church to Renew the Covenant," *Firmament* 2, no. 1 (1990): 5–9; *The Environment and the Christian* (Grand Rapids: Baker, 1991), 156.

3. This in turn is expected to have consequences for sea level and agriculture. See James G. Titus, "Effect of Climate Change on Sea Level Rise and the Implications for World Agriculture," *HortScience* 25, no. 12 (1990): 1567–72. See also Richard A. Houghton and George M. Woodwell, "Global Climatic Change," *Scientific American* 260, no. 4 (1989): 36–44; H. H. Lamb, "The Role of Atmosphere and Oceans in Relation to Climatic Changes and the Growth of Ice Sheets on Land," in *Problems in Paleoclimatology* (London: Interscience, 1964).

4. See Johan Moan, "Ozone Holes and Biological Consequences," *Journal of Photochemistry* 9, no. 2 (1991): 244–47, and F. S. Rowland, "Chlorofluorocarbons and the Depletion of Stratospheric Ozone," *American Scientist* 77 (1989): 36–45. The second article is a good summary of knowledge on the topic; a similar article is F. S. Rowland, "Chlorofluorocarbons, Stratospheric Ozone, and the Antarctic 'Ozone Hole,'" *Environmental Conservation* 15, no. 2 (1988): 101–15; an early article on the topic is M. J. Molina and F. S. Rowland, "Stratospheric Sink for Chlorofluoromethanes: Chlorine and Atom-Catalyzed Destruction of Ozone, *Nature* 249 (1974): 810–12.

5. See C. Sagan, O. B. Toon, and J. B. Pollack, "Anthropogenic Albedo Changes and the Earth's Climate," *Science* 206 (1979): 1363–68; Peter Usher, "Special Report: World Conference on the Changing Atmosphere: Implications for Global Security," *Environment* 1, no. 31 (1989): 25–35.

6. See C. J. Ganzer et al., "Evaluation of Soil Loss After One Hundred Years of Soil and Crop Management," *Agronomy Journal* 83 (1991): 74–77; J. A. Sandor and N. S. Eash, "Significance of Ancient Agricultural Soils for Long-Term Agronomic Studies and Sustainable Agricultural Research," *Agronomy Journal* 83 (1991): 29–37.

7. See David Pimentel et al., "Land Degradation: Effects on Food and Energy Resources," *Science* 194 (1976): 149–55; National Academy of Sciences, *Soil Conservation*, 2 vols. (Washington, D.C.: National Academy Press, 1986).

8. See Norman Myers, "Tropical Deforestation: The Latest Situation," *BioScience* 41, no. 5 (1991): 282; Allison G. Cook and W. Ted Hinds, *Environmental Conservation* 17, no. 3 (1990): 201–12; "The Forest Decline Enigma: What Underlies Extensive Dieback on Two Continents?" *BioScience* 37, no. 8: 542–46.

9. See David W. Moody, "Groundwater Contamination in the United States," *Journal of Soil and Water Conservation* (March–April 1990): 170–79.

Chapter 2: Religion and the Environment

1. Committee on Scientific Issues in the Endangered Species Act, *Science and the Endangered Species Act* (Washington, D.C.: Board on Environmental Studies and Toxicology, National Research Council, 1995). The project that is the subject of this report was approved by the Governing Board of the National Research Council, whose members are drawn from the councils of the National Academy of Sciences, the National Academy of Engineering, and the Institute of Medicine.

2. Ibid., 4–5, 25.

3. In a news release, they describe this bill introduced by Representatives Don Young (R-Ark.) and Richard Pombo (R-Calif.) as "A Disaster for Endangered Marine Wildlife." Their release was produced by Robert Irvin, deputy vice president for Marine Wildlife and Fisheries Conservation, Center for Marine Conservation, (202) 857-5551, irvinb@dccmc.mhs.compuserve.com.

4. Douglas John Hall, *Imaging God: Dominion as Stewardship* (Grand Rapids: Eerdmans, 1986), 8.

5. Ibid.

6. See, for example, J. Baird Callicott, "Genesis and John Muir," in Carol S. Robb and Carl J. Casebolt, eds., *Covenant for a New Creation: Ethics, Religion, and Public Policy* (Maryknoll, N.Y.: Orbis, 1991), 37; Max Oelschlaeger, *Caring for Creation* (New Haven: Yale University Press, 1994).

7. Tuan quote from David N. Livingstone, "The Historical Roots of Our Ecological Crisis: A Reassessment" (unpublished paper presented to Christianity Today Institute, Chicago, October 1993).

8. Huston Smith, *The Religions of Man* (New York: Harper & Row, 1958).

9. John Maynard Keynes, *Essays in Persuasion* (New York: W. W. Norton, 1963 [1930]), 371–72.

10. Charles Schultze, "The Public Use of Private Interest," *Harper's* (May 1977): 45–46. Schultze once was chairman of the President's Council of Economic Advisors.

11. And, of course, reading the passages on birds and flowers referred to here (Matt. 6:28 and Luke 12:27) assures us that if this be the case for birds and flowers, so much more is it the case for people! As the hymn puts it, "God will take care of you."

12. *The Kuyper Lecture*, Center for Public Justice, 6.

13. Joseph Sittler, "The Care of the Earth," in Franklin H. Littell, ed., *Sermons to Intellectuals* (New York: Macmillan, 1963). Sittler also reminds us, "There is an economics of joy; it moves toward the intelligence of use and the enhancement of joy. That this vision involves a radical new understanding of the clean and fruitful earth is certainly so. But this vision, deeply religious in its genesis, is not so very absurd now that natural damnation is in orbit, and man's befouling of his ancient home has spread his death and dirt among the stars."

14. These premises are adapted from the work of economist Eugene Dykema in Loren Wilkinson, *Earthkeeping in the Nineties: Stewardship of Creation* (Grand Rapids: Eerdmans, 1991), 242–46.

15. Gordon J. Spykman, *Reformational Theology: A New Paradigm for Doing Dogmatics* (Grand Rapids: Eerdmans, 1992), 140. In so writing, he references Kuyper, *Lectures on Calvinism*, 79.

16. Louis Berkhof, *Systematic Theology* (Grand Rapids: Eerdmans, 1941), 129. Berkhof also writes, "Since the Father takes the initiative in the work of creation, it is often ascribed to him economically."

17. This definition, interestingly, is the first definition of *economy* in *Webster's Third New International Dictionary of the English Language Unabridged* (Springfield, Mass.: Merriam-Webster, 1981), 720.

18. Donald Worster, *Nature's Economy: The Roots of Ecology* (San Francisco: Sierra Club, 1979), 37–38.

19. Ibid.

20. *Young's Literal Translation of the Holy Bible: A Revised Edition* (Grand Rapids: Baker, 1953).

21. Genesis 2:15 as interpreted by John Calvin, *Commentaries on the First Book of Moses, Called Genesis,* translated from the original Latin (1554), and compared with the French edition, by John King, vol. 1 (Grand Rapids: Eerdmans, 1948), 125.

22. *Custody* is from the Latin *custodia,* meaning "guarding, keeping," according to *Webster's Third New International Dictionary.*

23. This teaching is reinforced strongly by Revelation 11:18: "The time has come for . . . destroying those who destroy the earth."

24. See Ronald Manahan, "Christ as the Second Adam," in C. B. DeWitt, ed., *The Environment and the Christian: What Can We Learn from the New Testament?* (Grand Rapids: Baker, 1991), 45–56. "The work of the last Adam is as broad as the reach of the damage of the first Adam" (p. 55).

25. See, for example, Steven Shapin, *A Social History of Truth: Civility and Science in Seventeenth-Century England* (Chicago: University of Chicago Press), xv–xxiii, 126–92, 409–17.

26. See Richard Mouw, *Uncommon Decency: Christian Civility in an Uncivil World* (Downers Grove, Ill.: InterVarsity Press, 1992) and Ronald Manahan, "A Reexamination of the Cultural Mandate: An Analysis and Evaluation of the Dominion Materials" (Ph.D. diss., Grace Theological Seminary, 1985).

27. Francis Schaeffer, *Pollution and the Death of Man: The Christian View of Ecology* (Wheaton: Tyndale House, 1970), 81. I have split this quotation into two parts.

28. See M. Douglas Meeks, *God the Economist: The Doctrine of God and Political Economy* (Minneapolis: Fortress Press, 1989) and John Reumann, *Stewardship and the Economy of God* (Grand Rapids: Eerdmans, 1992).

29. As this is described by Herman Helmholtz, for example, in *On the Sensations of Tone as a Physiological Basis for the Theory of Music,* translated from the 4th German edition of 1877 (New York: Dover, 1954), especially 152–233.

Chapter 3: The Three Big Questions

1. Kuyper, "So God Loved the World!" 70–71.

2. Oliver O'Donovan, *Resurrection and Moral Order: An Outline for Evangelical Ethics* (Leicester, England: InterVarsity Press; Grand Rapids: Eerdmans, 1986), 13–14.

3. See Psalm 19:1, Acts 14:17, and Romans 1:20 for passages describing creation's testimony to God. Consider too that creation's telling of God's glory and love is echoed by Scripture. God lovingly provides rains, cyclings of water, and food for the creatures, fills people's hearts with joy, and satisfies the earth (Ps. 104:10–18; Acts 14:17). It is through this manifest love and wisdom that creation declares God's glory and proclaims the work of the Creator's hands (Ps. 19:1). Creation's evangelical testimony is so powerful that it leaves everyone without excuse (Rom. 1:20).

4. Quote from Loren Wilkinson, ed., *Earthkeeping in the Nineties: Stewardship of Creation* (Grand Rapids: Eerdmans, 1991), 299. For evidence that all Christian thought does not make this unbiblical distinction between the physical and the spiritual and does not espouse a salvation that turns people away from creation, see Wilkinson, *Earthkeeping,* 202–306. For a theological study of the importance of matter and of creation, and of the unbiblical hatred of creation by Marcion and Greek and Gnostic leaders, see Raymond C. Van Leeuwen, "Christ's Resurrection and the Creation's Vindication," in C. B. DeWitt, ed., *The Environment and the Christian:*

What Can We Learn from the New Testament? (Grand Rapids: Eerdmans, 1991), 57–71.

5. See "Theology, Science, and Creation: Extending the Horizon of Science and the Christian Faith," *Faculty Dialogue* 1995, no. 24, http://www.idnet.org/pub/facdialogue/24/dewitt24, for other reasons why it cannot be interpreted as "anything goes."

6. Creation here, as elsewhere, must be understood in the biblical sense of comprising all things created, including human beings.

7. For a more detailed treatment, see "Theology, Science, and Creation."

8. For a description of Noah's faithfulness in relation to this, see C. B. DeWitt, "The Price of Gopher Wood," *Faculty Dialogue*, no. 12 (1989): 59–62.

9. This paragraph is based on "Theology, Science, and Creation."

10. This paragraph is based on "Theology, Science, and Creation" and the next two paragraphs are taken from it directly.

11. That "care and keeping" were recognized before our time is evident in a prayer published in 1566: "Finally, O Lord, wilt Thou take us and our dear ones and all that concern us into Thy care and keeping" (from "A Prayer for All the Needs of Christendom," from the *Psalter* by Petrus Dathenus published in translation in the *Psalter Hymnal* [Grand Rapids: Board of Publications of the Christian Reformed Church, 1976], 183). That the connection is made in this prayer with Numbers 6:24 is indicated in its conclusion, which is a recitation of the Aaronic Blessing: "Jehovah bless thee, and keep thee; Jehovah make his face to shine upon thee, and be gracious unto thee; Jehovah lift up his countenance upon thee, and give thee peace. Amen."

12. This paragraph is taken from my piece to be published in the winter 1997 issue of the *Christian Research Journal*.

Chapter 4: Can We Afford to Keep the Lord's Behemoth?

1. There is remarkably fine material by Gordon Spykman here that goes beyond the constraints of this paper but that is highly relevant. See Spykman, *Reformational Theology*.

2. See Jeremiah 5:21–23 and 8:7 for more on boundaries and limits.

3. Sabbath observance brings a blessing by preventing people from exploiting themselves and the people and animals under their care, keeping them from hitting the absolute limit of seven days of work each week and helping them become more of what God wants them to be. Similarly, for the rest of creation, sabbath for the land limits creation's exploitation, bringing it God's blessing.

4. Abraham Kuyper, *Lectures on Calvinism*, 30.

Conclusion

1. Psalm 147, as versified in the *Psalter Hymnal* (Grand Rapids: Publication Committee of the Christian Reformed Church, Inc., 1959), 303.

2. Rendition by Paulos Mar Gregorios, "New Testament Foundations for Understanding the Creation," in Wesley Granberg-Michaelson, ed., *Tending the Garden: Essays on the Gospel and the Earth* (Grand Rapids: Eerdmans, 1987), 84 ("with neck outstretched"), and as presented at the Au Sable Forum ("on tippy toes") prior to its publication.

3. Kuyper, "So God Loved the World!" 70–71.

Response by Richard A. Baer Jr.

1. Aldo Leopold, *A Sand Country Almanac* (New York: Oxford University Press, 1949), 224–25.

Response by Thomas Sieger Derr

1. Holmes Rolston, *Environmental Ethics: Duties to and Values in the Natural World* (Philadelphia: Temple University Press, 1988), 103.

For nearly two decades Calvin B. DeWitt has been challenging Christian leaders to affirm the church's role in caring for the environment. He is professor of environmental studies at the University of Wisconsin at Madison and the director of the Au Sable Institute of Environmental Studies in Mancelona, Michigan. At the Au Sable Institute, environmental science majors from one hundred colleges and universities take highly specialized courses. Dr. DeWitt is the author, coauthor, or editor of *Earthkeeping in the Nineties: Stewardship of Creation* (Grand Rapids: Eerdmans, 1990), *The Environment and the Christian* (Grand Rapids: Baker, 1991), *Missionary Earthkeeping* (Macon, Ga.: Mercer University Press, 1992), and *Earthwise: A Biblical Response to Environmental Issues* (Grand Rapids: CRC Publications, 1994).

.